William R. Warner

Physicians' Complete Formulae Book

William R. Warner

Physicians' Complete Formulae Book

ISBN/EAN: 9783742818775

Manufactured in Europe, USA, Canada, Australia, Japa

Cover: Foto ©ninafisch / pixelio.de

Manufactured and distributed by brebook publishing software
(www.brebook.com)

William R. Warner

Physicians' Complete Formulae Book

PHYSICIANS'

COMPLETE

FORMULÆ BOOK

GIVING THE DOSES AND MEDICAL PROPERTIES,

—OF—

WM. R. WARNER & CO.'S

RELIABLE SOLUBLE-COATED

PILLS,

GRANULES and PARVULES

EFFERVESCENT SALTS, etc.

WITH REFERENCE NOTES RESPECTING
POISONS AND ANTIDOTES,
DISEASES AND REMEDIES,
Etc., Etc., Etc.

PUBLISHED BY

WILLIAM R. WARNER & CO.

PHILADELPHIA, NEW YORK AND LONDON.

1892.

CONTENTS

WARNER & CO.'S

PIL:

Cascara Cathartic

(DR. HINKLE.)

Each containing

Cascarin.	Ext. Belladon. $\frac{1}{8}$ gr.
Aloin, āā $\frac{1}{4}$ gr.	Strychnin, 1-60 gr.
Podophyllin, 1-6 gr.	Gingerine, $\frac{1}{8}$ gr.

Dose.—1 to 2 Pills.

This pill affords a brisk and easy cathartic, efficient in action and usually not attended with unpleasant pains in the bowels.

It acts mildly upon the liver, (Podophyllin), increases Peristalsis, (Belladonna), while the carminative effect of the Gingerine aids in producing the desired result, thus securing the most efficient and pleasant cathartic in use.

Per 100, 80 cts.

PREPARED BY

WM. R. WARNER & CO.

CHEMISTS,

1228 Market Street, Philadelphia.

3

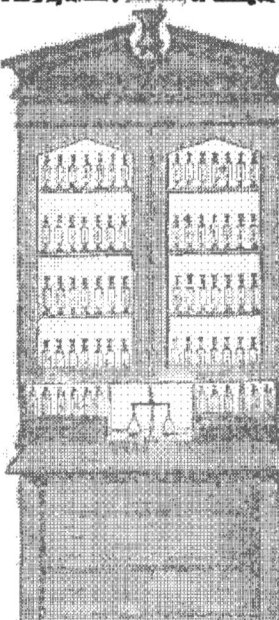

4

WARNER & CO.'S
Soluble Coated Pills.

These Pills are unsurpassed in their medicinal qualities, as only the best materials enter into their composition, while the most scrupulous care is exercised in their manufacture.

An extended laboratory experience comprising unceasing attention to details during a period of more than a quarter of a century enabled us to arrive at results otherwise unattainable.

We claim a method of coating which remains permanent and avoids the necessity for drying the mass so hard as to render it insoluble or inert. This scientific method which we do not hesitate to call our own is fully recognized and appreciated, as is demonstrated by the confidence reposed and the success attained.

It is our wish to emphatically impress upon the minds of prescribers, that our make of Pills will produce the effects as should be expected in connection with the drug employed, and that every other desire is subordinate to this end. We thank the Profession for the very liberal endorsement and patronage that has been accorded us and offer our assurance that our efforts shall, as heretofore, be directed towards the production of the highest class of Pharmaceutical preparations.

Respectfully,

WM. R. WARNER & CO.

TO THE MEDICAL PROFESSION.

While presenting this revised edition of our Formulary we take the opportunity to extend our thanks to our many friends; who have aided us by their patronage and influence in distributing Pharmaceutical Preparations of the highest standard and to express the hope that our efforts, in this direction, may continue to gain for us a confidence and feeling that our experience covering as it has, a period of thirty years and comprising deep scientific research and skillful manipulations, in conjunction with a conscientious desire to produce only the best, should at least warrant us in claiming a perfection of manufacture unexcelled.

Such facts are worthy the attention of Practitioners who are called upon to battle with disease, and who must in using ready prepared medicines, depend upon the correctness of manufacture to gain the looked for therapeutical effects, whereby their own reputations are sustained and the ills of their patients alleviated.

It may be well, in this connection, to reiterate that our endeavors have always been directed towards the production of first-class preparations, regardless of cost and passing the question of extreme cheapness which so often arises through competition and an anxiety to dispose of products; all of which very naturally has a tendency to depreciate quality correspondingly with the falling off in price.

We maintain that quality is of primary importance and in this we flatter ourselves that we have received the undivided encouragement and support of an intelligent profession, throughout the course of our business career.

With these few preliminary words outlining our policy, we beg leave to suggest the importance of specifying Warner & Co.'s make when ordering or prescribing.

Very Respectfully,

WM. R. WARNER & CO.

A COMPLETE FORMULÆ BOOK.

—OF—

WM. R. WARNER & CO.'S

RELIABLE AND PERMANENT

SOLUBLE COATED

PILLS.

Of the United States Pharmacopœia and Recipes
of Eminent Physicians.

PILLS.	BOTTLE.	
	100	500
Abernethy's (Aperient)..............	75	3 50
Pulv. Aloes Socot., 2 grs.		
Pulv. Ipecac., 5-6 gr.		
Pil. Hydrarg., 1 gr.		
Ext. Hyoscyam., 2 grs.		
Acid Arsenious. 1-20, 1-30, 1-50	40	1 75
and 1-60 grs..............................		
Medical properties—Anti-periodic, Al-		
terative. Dose, 1 to 2.	75	3 50
Aconitia. 1-60 gr....................		
Med. prop.—Nerve Sedative. Dose, 1.	75	3 50
Ague..................................		
Med. prop.—Anti-periodic. Dose, 2 to 4.		
Chinoidin, 2 grs.		
Ext. Coloc. Comp. ¼ gr.		
Ol. Pip. Nig., ⅙ gr.		
Ferri Sul., ½ gr.		
Aloes, U. S. P.......................	40	1 75
Med. prop.—Stimulating, Purgative.		
Dose, 1 to 3.		
Aloes Compound. (See Pil. Gen-		
tian Compound)......................	40	1 75
Med. prop.—Tonic, Purgative. Dose,		
2 to 4.		
Aloes et Asafœtida, U. S. P.....	40	1 75
Med. prop.—Purgative, Anti-spasmod-		
ic. Dose, 2 to 5.		

7

PILLS.	BOTTLE.	
	100	500
Aloes et Ferri.................	40	1 75
Med. prop.—Tonic, Purgative. Dose, 1 to 3.		
Pv. Aloes Socot., ½ gr.		
Pv. Zingiber Jam., 1 gr.		
Ferri Sul. Exs., 1 gr.		
Ext. Conii, ½ gr.		
Aloes et Ferri, U. S. P.	40	1 75
Med. prop.—Tonic, Purgative Dose, 1 to 3.		
Pur. Aloes, 1 gr.		
Aromat. Powd., 1 gr.		
Ferri Sul. Exs., 1 gr.		
Confect. Rose, q. s.		
Aloes et Mastich. (Lady Webster).	50	2 25
Med. prop.—Stimulating Purgative. Dose, 1 to 2.		
Aloes et Myrrhæ, U. S. P.......	50	2 25
Med. prop.—Cathartic, Emmenagogue. Dose, 3 to 6.		
Aloes et Nuc Vomicæ...........	50	2 25
Med. prop.—Tonic, Purgative. Dose, 1 to 2.		
Pulv. Aloes Soc., 1½ gr.		
Ext. Nuc. Vom., ½ gr.		
Aloin. ½ gr...............	1 00	4 75
Med prop.—Laxative.		
Aloin Comp................	50	2 25
Med. prop.—Tonic, Laxative. Dose 1 to 2.		
Aloin, ½ gr.		
Ext. Belladon. ¼ gr.		
Podophyllin, ⅛ gr.		
Aloin et Strychnin.............	60	2 75
Med. prop.—Tonic, Laxative.		
Aloin, 1·5 gr.		
Strychnine, 1-60 gr.		
Aloin et Strychnin. et Belladon.............	60	2 75
Used very largely and with great success in the treatment of habitual constipation.		
Med. prop.—Tonic, Laxative.		
Aloin, 1·5 gr.		
Strychnine, 1-60 gr.		
Ext. Belladon., ¼ gr.		

8

PILLS.	BOTTLE.	
	100	500
Aloin et Strychnin. et Bellad. Comp............,.....	75	3 50
Med. prop.		
Aloin, 1-5 gr.		
Ext. Belladon, ⅛ gr.		
Strychninæ, 1-60 gr.		
Ext. Cascara Sag., ½ gr.		
Alterative........................	50	2 25
Med. prop.—Alterative, with tendency to Mercurial Impression. Dose, 1 to 2.		
Mass. Hydrarg., 1 gr.		
Pulv. Opii., ⅛ gr.		
Pulv. Ipecac., ¼ gr.		
Alterative. (Dr. C. C. Cox).....	50	2 25
Med. prop.—Alterative, with tendency to Mercurial Impression. Dose, 1 to 2.		
Mass. Hydrarg., 1 gr.		
Pulv. Rhei., 1 gr.		
Sodii. Bicarb., 1 gr.		
Ammon. Bromid. 1 gr..........	75	3 50
Med. prop.—Sedative, Alterative, Resolvent. Dose, 1.		
Analeptic.............................	60	2 75
Med. prop.—Stimulant, Diaphoretic. Dose, 1 to 4.		
Pv. Antimonialis, ¾ gr.		
Pv. Res. Guaiac., 1 gr.		
Pv. Aloes Socot., ¾ gr.		
Pv. Myrrh., ½ gr.		
Anderson's Scot's...............	40	1 75
Med. prop.—Cathartic. 2 to 5.		
Pv. Aloes Socot.,		
Pv. Saponis.		
Pv. Colocynth.		
Pv. Gambogiæ.		
Anodyne........................	75	3 50
Med. prop.—Anodyne. Dose, 2 to 5.		
Pv. Camphoræ, 1 gr.		
Morphin Acetas, 1-20 gr.		
Ext. Hyoscyami, 1 gr.		
Ol. Res. Capsici, 1-20 gr.		
Anthelmintic......................	1 00	4 75
Med. prop.—Anthelmintic. Dose, 1 to 2.		
Santonin, 1 gr.		
Calomel, 1 gr.		

PILLS.

	BOTTLE.	
	100	500

Anti-Bilious (Vegetable)............ 50 | 2 25
Med. prop.—Cholagogue Cathartic.
Dose, 2 to 3.
Pv. Ext. Col. Co., 2½ grs. }
Podophyllin, ¼ gr. }

Anti-Chill........................ 1 00 | 4 75
Med. prop.—Anti-periodic. Applicable
to obstinate intermittents. Dose,
1 to 2.
Chinoidin, 1 gr. }
Ferri Ferrocyanid, 1 gr. }
Ol. Piper. Nig., 1 gr. }
Ac. Arsenious, 1-20 gr. }

Anti-Chlorotic................... 75 | 3 50
Med. prop.—Anti-chlorotic. Dose 1 to 2
Potass. Chlor., 1 gr. }
Ferri Chlor., ½ gr. }
Pv. Podophylli., 1 gr. }
Pv. Myrrhæ, ½ gr. }

Anti-Choromania.............. 75 | 3 50
Med. prop.—Anti-spasmodic. Dose, 1
to 2.
Zinci Valer, 2 grs. }
Ferri Valer, ¼ gr. }
Ext. Sumbul., ½ gr. }

Anti-Constipation............... 75 | 3 50
Dose, 1 to 4.
Podophyllin, 1-10 gr. }
Ext. Nuc. Vom. ⅓ gr. }
Pv. Capsici, ¼ gr. }
Ext. Belladon., 1-10 gr. }
Ext. Hyoscyami, ¼ gr. }

Anti-Constipation. (Palmer.)... 75 | 3 50
Dose, 1 to 2.
Aloes Soc't 1 gr. }
Ext. Hyoscyami, 1 gr. }
Ext. Nuc Vom., ⅓ gr. }
Ipecac Pulv., 1-10 gr. }

Anti-Dyspeptic.................. 1 00 | 4 75
Med. prop.—Applicable where Debility
and Impaired Digestion exist. Dose,
1 to 2.
Strychnine, 1-40 gr. }
Ext. Belladon, 1-10 gr. }
Pulv. Ipecac., 1-10 gr. }
Mass. Hydrarg., 2 grs. }
Ext. Coloc. Comp. 2 grs. }

10

PILLS.	BOTTLE.	
	100	500
Anti-Dyspeptic. (Fothergill)....	75	3 50
Dose, 1.		
Pv. Ipecac., ¾ gr.		
Strychnine, 1-20 gr.		
Pv. Piper. Nig. 1½ gr.		
Ext. Gentian, 1 gr.		
Anti-Malarial...............	1 25	6 00
Med. prop.—Anti-malarial. Dose, 1 to 2		
Quininæ Sulph., 1 gr.		
Cinchoninæ Sulph., ⅓ gr.		
Ferri Sulph. Exs. ¼ gr.		
Ac. Arsenious, 1-40 gr.		
Anti-Malarial (Philadelphia).....	1 00	4 75
Med. prop.—Anti-malarial. Dose, 1 to 2.		
Ferri Sulph., 1 gr.		
Pv. Capsici, ⅛ gr.		
Cinchonid. Sulph., 2 grs.		
Strychnine, 1-30 gr.		
Anti-Malarial. (McCaw).........	1 50	7 25
Dose, 1 to 2.		
Quininæ Sulph., 1 gr.		
Ferri Sulph. Exs., ¼ gr.		
Ac. Arsenious, 1-80 gr.		
Gelsemin, ¼ gr.		
Podophyllin, ⅔ gr.		
Ol. Res. Pip. Nig., 1-16 gr.		
Antimonii Comp., U.S. P. (Pil.		
Plummer)........................	40	1 75
Med. prop.—Alterative. Dose, 1 to 3.		
Anti-Periodic...................	80	3 75
Med. prop.—Anti-periodic. Dose 1		
to 3.		
Cinchonidin Sulph., 1 gr.		
Res. Podophylli, 1-20 gr.		
Strychninæ Sul., 1-33 gr.		
Gelsemin., 1-20 gr.		
Ferri Sulph. Exs., ½ gr.		
Ol. Res. Capsici, 1-10 gtt.		
Anti-Periodic. (One-half size).....	45	2 00
Dose, 1 to 3.		
Anti-Spasmodic.................	75	3 50
Med. prop.—Anti-spasmodic. Dose, 1		
to 2.		
Ext. Hyoscyami, ½ gr.		
Morphinæ Acetas., 1-10 gr.		
Brom. Camphor., ¼ gr.		
Pv. Capsici, ⅔ gr.		

11

PILLS.	BOTTLE.	
	100	500
Antisplenetie....................	60	2 75
Med. prop.—Anti-splenetic. Dose, 2 to 4.		
Pv. Aloes Soc., 1 gr.		
Pv. Ammoniaci, ¼ gr.		
Pv. Myrrhæ, ½ gr.		
Ext. Bryony, 1 gr.		
Aperient........................	85	4 00
Med. prop.—Aperient, Tonic. Dose, 1 to 2.		
Ext. Nuc. Vom., ½ gr.		
Ext. Hyoscyami, ¼ gr.		
Ext. Coloc. Co., 2 grs.		
Aperient. (Dr. Fordyce Barker.)...	1 00	4 75
Med. prop.—Aperient.		
Ext. Coloc. Co., 1⅔ gr.		
Ext. Nuc. Vom., ½ gr.		
Ext. Hyoscyam., 1¼ gr.		
Pulv. Ipecac, 1-12 gr.		
Pulv. Aloes Soc't, 5-12 gr.		
Res. Podophylli., 1-12 gr.		
Aperient. (Drysdale.)............	60	2 75
Med. prop.—Aperient. Dose, 1 to 2.		
Pv. Rhei, 1¼ gr.		
Pv. Ipecac, 5-12 gr.		
Pv. Aloes Soc., 1¼ gr.		
Pv. Nuc. Vom., ½ gr.		
Arthrosia. (See special page)......	80	
Med. prop.—Antilithic, Tonic, Alterative.		
Acid Salicylic.		
Ext. Colchicum.		
Ext. Phytolacca.		
Res. Podoyhylli.		
Quinine.		
Pv. Capsicum.		
Asafœtida, U. S. P............	40	1 75
Med. prop.—Nerve Stimulant. Dose, 1 to 3.		
Asafœtida. 2 grs................	40	1 75
Med. prop.—Nerve Stimulant. Dose, 2 to 4.		
Asafœtida Comp...............	40	1 75
Med. prop.—Tonic and Nerve Stimulant. Dose, 2 to 5.		
Asafœtidæ, 2 grs.		
Ferri Sulph. Exsic., 1 gr.		

12

PILLS.	BOTTLE.	
	100	500
Asafœtida et Aloes.............	40	1 75
Med. prop.–Purgative, Anti-spasmodic.		
Dose, 2 to 5.		
Asafœtida et Rhei...............	75	3 50
Med. prop.—Tonic, Laxative, Nerve		
Stimulant. Dose, 2 to 4.		
Asafœtidæ, 1 gr.)		
Pulv. Rhei, 1 gr. }-		
Ferri Redact., 1 gr.)		
Astringent.....................	60	2 75
Med. prop.—Astringent. Dose, 1 to 2.		
Ext. Geranii, 2 grs.)		
Pv. Opii, ¼ gr.		
Ol. Menth. Pip., 1-20 gtt. }		
Ol. Res. Zingiber., 1-20 gtt.)		
Atropina. 1-6 and 1-100 gr.........	75	3 50
Med. prop.—Anodyne.		
Atropinæ Sulph. 1-60 gr.........	75	3 50
Med. prop.—Anodyne.		
Bismuth Subcarb. 3 grs.........	75	3 50
Med. prop.—Sedative. Dose, 2 to 5.		
Bismuth Subnit. 3 grs..........	75	3 50
Med. prop.—Sedative, Anti-periodic.		
Dose, 1 to 5.		
Bismuth et Ext. Nuc. Vom......	1 50	7 25
Med. prop.—Sedative, Tonic. Dose,		
1 to 2.		
Bismuth Subcarb., 4 grs.)		
Ext. Nuc. Vomicæ, ¼ gr.)		
Bismuth et Ignatia............	1 50	7 25
Med. prop.—Sedative, Anti-periodic,		
Tonic. Dose, 1 to 2.		
Bismuth. Subcarb., 4 grs.)		
Ext. Ignatiæ Amara, ¼ gr.)		
Blennorrhagic. (W & Co.)........	1 00	4 75
Dose, 1 to 2.		
Terebinth Alb., 1½ grs.)		
Ext. Humuli, ¼ gr.		
Camph. Monobrom, ¾ gr.		
Res. Podophyllin, ⅛ gr.)		

The remedy *par excellence* for Chronic
Blennorrhœa, uncomplicated with organic
stricture, very frequently effecting a
speedy cure in gleet of long standing.
It is also almost equally serviceable as
a remedy for cystorrhœa and inflammation of whatever kind effecting the urinary
or sexual organs.

13

PILLS.	BOTTLE.	
	100	500

Caffein Citras. 1 gr.............. 2 75 | 13 50
Med. prop.—Nerve Stimulant. Dose, 1.

Calcium Sulphide. 1-10 gr....... 50 | 2 25
Med. prop.—Useful in Cutaneous Diseases. Dose, 2 to 4.

[Extract from Dr. Howard Cane's Article in the London Lancet.] From the great frequency of occurrence of acne, and from its manifesting itself on the faces of individuals of both sexes, any therapeutic agent which promises success in this often intractable skin disease will be welcomed by most practitioners. I do not bring the sulphide of calcium forward as a new remedy in the treatment of this disease, for it was recommended some years ago by Dr. Sydney Ringer, but I wish to bring it more prominently into notice as a drug which will often prove of signal service in acne when other means have failed. The success which I attained in my first case which was of a most obstinate nature, led me to try it in others.

Calcium Sulphide. ¼ gr........ 60 | 2 75
Med. prop.—Useful in Cutaneous Diseases. Dose, 2 to 3.

Calcium Sulphide. ½ gr........ 75 | 3 50
Med. prop.—Useful in Cutaneous Diseases. Dose, 2 to 3.

Calcium Sulphide. 1 gr......... 1 00 | 4 75
Med. prop.—Useful in Cutaneous Diseases. Dose, 1 to 2.

Calomel. ½ gr., 1, 2, and 3 grs...... 40 | 1 75
Med. prop.—Alterative. Dose, 1 to 3.

Calomel. 5 grs.................... 50 | 2 25
Med. prop.—Alterative, Purgative. Dose, 1 to 3.

**Calomel et Opii................. 85 | 4 00
Med. prop.—Cathartic, Anodyne. Dose 1
Calomel, ... 2 grs.
Opium, ... 1 gr.

**Calomel et Rhei................. 75 | 3 50
Med. prop.—Mild Purgative. Dose, 1 to 3.
Calomel, ... ⅓ gr.
Ext. Rhei, ... ½ gr.
Ext. Coloc. Comp., .. ½ gr.
Ext. Hyoscyami, ... ⅙ gr.

14

PILLS.	BOTTLE.	
	100	500
Camphor. ½ gr............... Med. prop.—Diaphoretic Carminative. Dose, 2 to 4.	40	1 75
Camphor. 1 gr............... Med. prop.—Diaphoretic Carminative. Dose, 1 to 2.	50	2 25
Camphor et Ext. Hyoscyamus. Med. Prop —Anodyne, Cerebral Stimulant. Dose, 1 to 2. Camphor, 1 gr } Ext. Hyoscyami, Eng., 1 gr }	50	2 25
Camphor Monobromated. 2 grs. Med. prop.—Anti-spasmodic. Dose, 1 to 2.	1 25	6 00
Cascara Comp. .... Med. prop —Laxative, Cathartic. Ext. Cascara Sagrad., 3 grs. } Res. Podophylli, ⅛ gr. }	75	3 50
Cathart. Comp., U.S.P. .... Med. prop.—Cathartic. Dose, 2 to 4. These pills are made strictly in accordance with the formula as directed by the Pharmacopœia.	30	1 35
Cathart. Comp., Imprv'd. .... Med. prop.—Cathartic. Dose, 2 to 4. Ext. Coloc. Comp. Ext. Jalap. Podophyllin, Leptandrin. Ext. Hyoscyami. Ext. Gentian. Ol. Menth. Pip.	30	1 35
Cathart. Comp., Vegetable ...... Med. prop.—Cathartic. Dose, 2 to 3. Podophyllin, Scammony. Ext Colocynth. Aloes, Soap, and Cardamom.	30	1 35
Cathart. Comp., Cholagogue ... Med. prop.—Cathartic Dose, 2 to 4 Res. Podophylli, ¼ gr. } Pil. Hydrarg., ¼ gr. } Ext. Hyoscyami, ⅛ gr. } Ext. Nuc. Vom., 1-16 gr. } Ol. Res. Capsici, ⅛ gtt. }	60	2 75
Caulophyllin. 1-10 gr............ Med. prop.—Emmenagogue. Dose, 1 to 3.	40	1 75

15

PILLS.	BOTTLE.	
	100	500

Chalybeate. 3 grs. 60
Med. prop.—Antichlorotic.

Ferri Sulph.,   1½ gr. }
Potassa Carb.,   1½ gr. }

This combination which we have successfully and scientifically put in pill form
produces when taken into the stomach,
Carbonate of Protoxide of Iron, (Ferrous
Carbonate) in a quickly assimilable condition.

This pill contributed to make the reputation of Niemeyer, and the following lan-
guage, which speaks without comment, is
taken from his TEXT BOOK ON THE
PRACTICE OF MEDICINE.

PROF. NIEMEYER says: "For more
than twenty years I have used these pills
almost exclusively in Chlorosis, and have
witnessed such brilliant results from them
in a large number of cases that I have
never needed any opportunity to experi-
ment with other articles. At Madgeburg
and Greifswald I often had to send my
recipe for the pills to a great distance, my
good fortune in the treatment of Chlorosis
—to which, by-the-by, I owe the rapid
growth of my practice—having given me
great repute as the possessor of a sover-
eign remedy against that disease."

The dose of Pil. Chalybeate is from 1 to
4 at meal times and is recommended and
successfully used in the treatment of
Pulmonary Phthisis or Consumption,
Anæmia and Chlorosis, Caries and Scrofulous Abscesses, Chronic Discharges,
Dyspepsia, Loss of Appetite, etc.

*The physician may see that he is
obtaining exactly what he prescribes, by
ordering in bottles containing 100 each.*

Ceril Oxalas. 1 gr . . . . 1 00 | 4 75
Med. prop.—Nerve Tonic. Dose, 1 to 3.

Chapman's Dinner Pills . . . . . . 60 | 2 75
Med. prop.—Stimulating, Laxative.
Dose, 1 to 3.
Pulv. Aloes Soc.
Pulv. Rhei Opt.
Gum Mastich.

Chinoidin. 1 gr. . . . . 40 | 1 75
Med. prop.—Tonic, Anti-periodic.
Dose, 2 to 4.

16

PILLS.	BOTTLE. 100	500
Chinoidin. 2 grs............... Med. prop.—Tonic, Anti-periodic. Dose, 2 to 4.	50	2 25
Chinoidin Comp............... Med. prop.—Tonic, Anti-periodic. Dose, 1 to 2. Chinoidin, 2 grs. Ferri Sulph. Exsic., 1 gr. Piperin, ½ gr.	75	3 50
Cholagogue. (Dr. Blackwood.).... Med. prop.—"An admirable Cholagogue." Dose, 1 to 2. Cinchonid. Sulph., ½ gr. Euonymin, ½ gr. Leptandrin, ½ gr. Iridin, ½ gr. Juglandin, ½ gr. Podophyllin, ⅙ gr. Ext. Belladon., ⅙ gr. Ext. Nuc. Vom., ⅙ gr. Ext. Hyoscyam., ⅙ gr.	1 00	4 75
Cimicifugin. 1-10 gr............... Med. prop.—Tonic, Nerve Stimulant. Dose, 1 to 4.	40	1 75
Cinchonine Sulph. 1½ gr...... Med. prop.—Tonic, Anti-periodic. Dose, 1 to 3.	60	2 75
Cinchonine Sulph. 2 grs........ Med. prop.—Tonic, Anti-periodic. Dose, 1 to 3.	60	2 75
Cinchonidin. Comp. (Warner & Co.)............... Med. prop.—Tonic, Anti-periodic. Cinchonid. Sulph., 2 grs. Ac. Salicylic., 1 gr. Pv. Opii, ½ gr. Ol. Res. Capsici., ¼ gr. This pill is also termed Pil. Salicylic Acid Comp.	1 50	7 25
Cinchonidine Salicylate. 2½ grs............... Med. prop.—Anti-rheumatic. Dose, 1 to 2.	1 50	7 25
Cinchonidine Sulph. 1 gr...... Med. prop.—Anti-malarial, Anti-periodic. Dose, 1 to 3.	40	1 75

17

PILLS.	BOTTLE. 100	500
Cinchonidinæ Sulph. 2 grs..... Med. prop.—Anti-malarial, Anti-perio-dic. Dose, 1 to 3.	40	1 75
Cinchonidinæ Sulph. 3 grs..... Med. prop.—Anti-malarial, Anti-perio-dic. Dose, 1 to 2.	50	2 25
Cinchonidinæ Sulph. 5 grs..... Med. prop.—Anti-periodic, Anti-mala-rial. Dose, 1 to 2.	1 00	4 75
Cincho-Quinine. 1 gr. .... Med. prop.—Tonic, Anti-periodic. Dose, 1 to 3.	1 00	4 75
Cincho-Quinine. 2 grs........... Med. prop.—Tonic, Anti-periodic. Dose, 1 to 2.	1 90	9 25
Coccia... .... Med. prop.—Hydragogue cathartic. Dose, 2 to 4. Pulv. Res. Scammon., 1 gr. Pulv. Soc. Aloes, 1¼ gr. Pulv. Colocynth., ⅛ gr. Potass. Sulph., ⅛ gr. Ol. Caryophyl., ⅛ gtt.	90	4 25
Codein. ¼ gr................ Med. prop.—Anodyne, replacing mor-phine without the usual disagreeable after-effects produced by the latter. Dose, 1 to 2.	1 25	6 00
Colocynthidis Comp., U. S. P. 3 grs.................... Med. prop.—Purgative. Dose, 2 to 5.	80	3 75
Colocynth. et Hydrarg. et Ipe-cac, ... Med. prop.—Cholagogue, Cathartic. Dose, 1 to 3. Pulv. Ext. Coloc. Co., 2 grs. Pil. Hydrarg., 2 grs. Pulv. Ipecac, ⅙ gr.	75	3 50
Colocynth. et Hyoscyamus. .. Med. prop.—Gentle laxative. Dose, 1 to 2. Ext. Coloc. Comp., 2½ grs. Ext. Hyoscyami, 1½ gr.	75	3 50

18

PILLS.	BOTTLE.	
	100	500

Cook's 3 Grs. .... 50 | 2 25
Med. prop.—Pergative. Dose, 2 to 4.

Pulv. Aloes Soc.,	1 gr.
Pulv. Rhei,	½ gr.
Calomel,	½ gr.
Sapon. Hispan.,	½ gr.

Copaibæ, U. S. P. .... 50 | 2 25
Med. prop.— Alterative to Mucous Membrane. Dose, 2 to 6.

Copaibæ et Ext. Cubebæ .... 80 | 3 75
Med. prop.—Alterative to Mucous Membrane. Dose, 2 to 4.

| Pil. Copaibæ, | 3 grs. |
| Oleo-Resin. Cubebæ, | 1 gr. |

Copaibæ Comp. .... 80 | 3 75
Med. prop. — Alterative to Mucous Membrane, Tonic. Dose, 2 to 4.

Pil. Copaib.
Resin. Guaiac.
Ferri Citras.
Oleo-Resin. Cubebæ.

Corrosive Sublimate. 1-12, 1-20, 1-40 and 1-100 gr. .... 40 | 1 75
Med. prop.—Mercurial, Alterative, Corrosive Sublimate has been administered with most gratifying results in certain forms of Chronic Dyspepsia. Ringer and other eminent therapeutists extol it very highly in 1-100 gr. doses in dysentery of children, regarding it as almost specific.

Damianæ cum Phos. et Nuc. Vom. 1 50
Med. prop.—Aphrodisiac. Dose, 1 to 2.

Ext. Damianæ,	2 grs.
Phosphori,	1-100 gr.
Ext. Nuc. Vom.,	⅛ gr.

A valuable remedy indicated in sexual debility, over work of the brain, impotency, etc. It is also highly recommended as an uterine tonic. Also of value in Leucorrhœa, Amenorrhœa, Dysmenorrhœa, etc.

Diaphoretic. .... 75 | 3 50
Med. prop.—Diaphoretic. Dose, 1 to 2.

Morphniæ Acetas.,	1-25 gr.
Pv. Ipecac,	¼ gr.
Pv. Potass. Nitras.,	1 gr.
Pv. Camphoræ,	¼ gr.

PILLS.	BOTTLE.	
	100	500

Digestiva 75

Med. prop.—Useful in Indigestion.
Dose, 1 to 2.

Pepsin Concentrat.,	1 gr.
Pv. Nuc. Vom.,	¼ gr.
Gingerine,	1-16 gr.
Sulphur,	⅛ gr.

This combination is very useful in relieving various forms of Dyspepsia and
Indigestion and will afford permanent
benefit in cases of enfeebled digestion,
where the gastric juices are not properly
secreted.

As a corrective of nausea or lack of
appetite in the morning, induced by over
indulgence in food or stimulants during the
night, these pills are unsurpassed; they
should be taken in doses of two pills before
retiring or in the morning at least one
hour before eating; the first mentioned
time is the most desirable as the effects
are more decided, owing to the longer
period for action and the natural rest is
more fully experienced through their mild
but effective influence.

As a dinner pill, Pil: Digestiva is unequalled and may be taken in doses of a
single pill either before or after eating.

Digitalin. (Alkaloid.) 1-60 gr...... 75 3 50

Med. prop.—Arterial sedative. Dose,
1 to 2.

Digitalis Comp. 50 2 25

Med. prop.—Arterial sedative. Dose,
1 to 3.

Pv, Digitalis,	1 gr.
Pv. Scillæ,	1 gr.
Potass. Nitras.,	2 grs.

Diuretic 50 2 25

Med. prop.—Diuretic, Antacid. Dose,
1 to 3.

Sapon. Hispan. Pv.,	2 grs.
Sodii Carb. Exsic.,	2 grs.
Ol. Baccæ Junip.,	1 gtt.

Dupuytren 50 2 25

Med. prop.—Specific Alterative. Dose,
1.

Pulv. Guaiac.,	3 grs.
Hydg. Chlor. Cor.,	1-10 gr.
Pulv. Opii,	⅛ gr.

20

PILLS.	BOTTLE.	
	100	500
Eccoprotic.................................	60	2 75
Med. prop.—Mild cathartic. Dose, 2		
to 4.		
Ext. Aloes Soc., 2 grs. }		
Ext. Nuc. Vomicæ, 1-5 gr. }		
Res. Podophylli, 3-10 gr. }		
Ol. Caryophyl., 1-10 gtt. }		
Elaterium. (Clutterbuck's) 1-10 gr.	95	4 50
Med. prop.—Diuretic, Hydragogue		
Cathartic. Dose, 1 to 2.		
Emmenagogue............................	1 25	6 00
Med. prop.—Active Emmenagogue,		
Tonic. Dose, 1 to 3.		
Ergotine, 1 gr. }		
Ext. Hellebor. Nig., 1 gr. }		
Aloes, 1 gr. }		
Ferri Sul. Exs., 1 gr. }		
Ol. Sabinæ, ½ gr. }		
Emmenagogue. (Mutter)..........	40	1 75
Med. prop.—Emmenagogue. Dose, 1		
to 3.		
Ferri Sulph. Exs., 1½ gr. }		
Aloes Pv., ½ gr. }		
Terebinth. Alb., 1¼ gr. }		
Ergotin. 1 gr...........................	1 00	4 75
Med. prop.—Parturient.		
Ergotin. 3 grs........................	1 50	7 25
Med. prop.—Parturient.		
Dose, 1 to 2.		
Ergotin. Comp. (Dr. Reeves.)....	1 75	8 50
Med. prop.—Sedative, Parturient.		
Ergotin, 3 grs. }		
Ext. Cannab. Ind., ¼ gr. }		
Ext. Belladon., ¼ gr. }		
Extract Belladonna. (Eng.) ¼ gr.	40	1 75
Med. prop.—Anodyne.		
Extract Cannabis Indica. ¼ gr.	60	2 75
Med. prop.—Anodyne.		
Ext. Guarana. 3 grs..............	2 00	9 75
Med. prop.—Nervine. Dose, 1 to 3.		
Extract Hyoscyamus. (Eng.)		
¼ gr...	40	1 75
Med. prop.—Nerve Stimulant.		
Ext. Coca. 3 grs.....................	80	3 75
Med. prop.—A powerful Tonic and		
Sedative. Dose, 1 to 2.		

PILLS.	BOTTLE.	
	100	500
Extract Ignatia Amara. ¼ gr.	50	2 25
Med. prop.—Nerve Sedative. Dose, 1 to 2.		
Extract Nuc. Vom. ¼ and ½ gr.	40	1 75
Med. prop.—Nerve Stimulant. Dose, 1 to 3.		
Fel Bovinum......................	50	2 25
Med. prop.—Laxative. Dose, 1 to 3.		
Fel Bovis. Ins. 2 grs. }		
Pv. Zingiber Jam., 1 gr. }		
Ferri Iodid. 1 gr...........	80	
Med. prop.—Tonic, Alterative. Dose, 1 to 2.		

In cases where Iodide of Iron is prescribed, it is absolutely necessary, for the physician, who relies on the therapeutic action for beneficial results, that the compound should be perfectly protected, and so prepared as to remain inalterable and stable.

With this important fact in view, we have devoted special study to Iodide of Iron in pilular form, and are warranted in announcing that Warner & Co.'s Iodide of Iron Pills meet all the requirements, and are the most perfect preparation of the kind.

A salt is formed and so prepared as to guard against oxidation, and will remain unchanged for years. A coating of pure sugar renders them pleasant to administer, and further insures protection.

In proof of the above statement, a pill cut through presents all the characteristics of a perfect pill mass and the presence of Iodide Iron, without the free Iodine, forming a clear solution; and dissolving *readily* if thrown into a glass of water.

The dose of Iodide Iron Pills is from ONE to TWO at meal time and is recommended and successfully used in the treatment of Pulmonary Phthisis or Consumption, Anæmia and Chlorosis, Caries and Scrofulous Abscesses, Chronic Discharges, Dyspepsia, Loss of Appetite, etc.

Ferri Carb. (Vallett's) **U. S. P.**		
3 grs..................	40	1 75
Med. prop.—Tonic. Dose, 1 to 4.		

22

PILLS.	BOTTLE.	
	100	500
Ferri Citras, U. S. P. 2 grs.....	50	2 25
Med. prop.—Tonic. Dose, 1 to 3.		
Ferri Comp., U. S. P.	40	1 75
Med. prop.—Tonic, Emmenagogue. Dose, 2 to 6.		
Ferri Lactas. 1 gr.................	50	2 25
Med. prop.—Tonic. Dose, 1 to 3.		
Ferri Pyrophos. 1 gr...........	40	1 75
Med. prop.—Tonic. Dose, 1 to 3.		
Ferri et Quas. et Nuc. Vom....	75	3 50
Med. prop.—Tonic, Nerve Stimulant. Dose, 1 to 2.		
Fer. per Hydrog., 1½ gr.		
Ext. Quassiæ, 1 gr.		
Ext. Nuc. Vom., ¼ gr.		
Pulv. Saponis, ½ gr.		
Ferri et Strychnin...............	75	3 50
Med. prop.—Tonic, Nerve stimulant. Dose, 1 to 2.		
Ferrum per Hydrog., 2 grs.		
Strychninæ, 1-60 gr.		
Ferri et Strychninæ Cit........	75	3 50
Med. prop.—Tonic, Nerve stimulant. Dose, 1 to 2.		
Strych. Citrac., 1-50 gr.		
Ferri Citras., 1 gr.		
Ferri Sulph. Exs. 2 grs..........	40	1 75
Med. prop.—Tonic. Dose, 2 to 4.		
Ferri Valer. 1 gr.................	1 00	4 75
Med. prop.—Tonic, Anti-spasmodic. Dose, 1 to 2.		
Ferrum. (Quevenne's.) 1 gr.......	50	2 25
Med. prop.—Tonic. Dose, 1 to 3.		
Ferrum. (Quevenne's.) 2 grs.......	75	3 50
Med. prop.—Tonic. Dose, 1 to 2.		
Galbani Comp..............	50	2 25
Med. prop.—Anti-spasmodic. Dose, 2 to 4.		
Galbani, 1½ gr.		
Pv. Myrrh., ½ gr.		
Asafœt, ½ gr.		
Gambogiæ Comp................	40	1 75
Med. prop.—Active purgative. Dose, 2 to 5.		
Pv. Gambogia.		
Pv. Aloes Socot.		
Pv. Zingib. Jam.		
Pv. Saponis. [23]		

PILLS.	BOTTLE.	
	100	500
Gelsemin. 1-16 gr................ Med. prop.—Arterial sedative. Dose, 1 to 4.	40	1 75
Gelsemin. ⅛ gr................. Med. prop.—Arterial sedative. Dose, 1 to 2.	50	2 25
Gelsemin. ¼ gr................. Med. prop.—Arterial Sedative. Dose, 1 to 2.	75	3 50
Gentian. Comp. (Aloe Comp.).... Med. prop.—Tonic, Purgative. Dose, 2 to 4. Ext. Gentian, ⅔ gr.⎫ Pv. Aloes Soc., 2 grs. ⎬ Ol. Carui, 1-5 gr.⎭	40	1 75
Gonorrhœa......................... Med. prop.—Tonic, Alterative to Mucous Membrane. Dose, 1 to 3. Pulv. Cubeb., 2 grs.⎫ Bals. Copaib. Solid., 1 gr.⎪ Ferri Sulph., ½ gr.⎬ Terebinth. Venet., 1½ gr.⎭	60	2 75
Heim's (Niemeyer.).............. Med. prop.—Anti-periodic, Tonic. Dose, 1. Quininæ Sulph., 1 gr.⎫ Pulv. Digital. Fol., ½ gr.⎪ Pulv. Ipecac, ¼ gr.⎬ Pulv. Opii, ¼ gr.⎭	1 25	6 00
Helonin. 1-10 gr.................. Med. prop.—Cathartic. Dose, 1 to 4.	50	2 25
Hepatica......................... Med. prop.—Cholagogue, Cathartic. Dose, 1 to 2. Pil. Hydrarg., 3 grs.⎫ Ext. Coloc. Comp., 1 gr.⎬ Ext. Hyoscyami, 1 gr.⎭	80	3 75
Hooper. (Female Pills.) 2½ grs.... Med. prop.—Emmenagogue. Dose, 1 to 3	40	1 75
Hydrarg. 5 grs.................... Med. prop.—Mercurial alterative. Dose, 1 to 2.	50	2 25
Hydrargyri, U. S. P. 3 grs...... Med. prop.—Mercurial alterative. Dose, 2 to 3.	40	1 75
Hydrarg. Bin Iodide. 1-16 gr... Med. prop.—Alterative. Dose, 1 to 3.	40	1 75

PILLS.	BOTTLE. 100	500
Hydrargyri Comp............	75	3 50
Med. prop.--Mercurial alterative. Dose, 1 to 2. Mass. Hydrarg., 1 gr.) Pulv. Opii, ½ gr. } Pulv. Ipecac, ¼ gr.)		
Hydrargyri Iod. et Opii. (Ricords.).............	75	3 50
Med. prop.--Mercurial alterative. Dose, 1 to 2. Hydrarg. Iodid., 1 gr.) Pulv. Opii, ⅓ gr.)		
Hydrarg. Prot Iodide. 1-5 gr..	40	1 75
Med. prop.--Alterative. Dose, 1 to 4. Our preparation of Prot Iodid. Mercury is made by precipitation and is entirely free from traces of the Bin-Iodid.		
Hydrarg. Prot Iodide. ⅛ gr...	40	1 75
Med. prop.--Alterative. Dose, 1 to 4.		
Hydrarg. Prot Iodide. ¼ gr...	40	1 75
Med. prop.--Alterative. Dose, 1 to 2.		
Hydrarg. Prot Iodide. ½ gr...	50	2 25
Med. prop.--Alterative. Dose, 1 to 2.		
Hydrastin. ½ gr............	95	4 50
Med. prop.--Cathartic. Dose, 1 to 2.		
Hyoscyamiæ. 1-100 gr..........	3 00	14 75
(Crystals, Pure Alkaloid.) Med. prop.--Anodyne, Soporific. Dose, 1.		
Iodoform. 1 gr...............	1 00	4 75
Med. prop,--Tonic, Alterative. Dose, 1 to 2.		
Iodoform et Ferri..............	1 50	7 25
Med. prop.--Tonic, Alterative. Dose, 1 to 2. Iodoform, 1 gr.) Ferri Redact., 1¼ gr.)		
Iodoform et Ferri et Nuc. Vom	1 50	7 25
Med. prop.--Tonic, Alterative. Dose, 1 to 2. Iodoform, 1 gr.) Ferri Redact., 1 gr. } Ext. Nuc. Vom., ¼ gr.)		

PILLS.	BOTTLE.	
	100	500
Iodoform et Hydrarg..........	1 50	7 25
Med. prop.—Alterative. Dose, 1 to 3.		
Iodoform, ½ gr.		
Hydrarg. Prot Iodid., ⅛ gr.		
Iodoform et Nuc. Vom. Comp...	1 50	7 25
Med. prop.—Alterative, Tonic, Laxa-		
tive, Repellant. Dose, 1 to 3.		
Iodoform ¼ gr.		
Ext. Nuc. Vom., ⅛ gr.		
Podophyllin, 1-16 gr.		
Ext. Belladon., ⅛ gr.		
Iodoform et Quinine............	1 25	6 00
Med. prop.—Alterative, Tonic. Dose,		
1 to 3.		
Iodoform, ½ gr.		
Quininæ Bisulph., ½ gr.		
Iodoform et Quinina et Ferri.	1 75	8 50
Med. prop.—Tonic, Alterative. Dose,		
1 to 2.		
Iodoform, 1 gr.		
Ferri Carb., (Vallet's) 2 grs.		
Quininæ Sul., ½ gr.		

Iodoform therapeutically is alterative, nervine, sorbefacient, anti-periodic, and anæsthetic. As an alterative it acts with more rapidity than other medicines of that class, in doses of one, two, or three grains, repeated thrice daily. As a nervine it is prompt and efficient; while it gives nervous strength, it calms speedily the most severe pains.

It is rapidly absorbed into the blood.

Accumulative effects have not been observed.

Iodoform is destitute of any local Irritant action, and has that advantage over all other iodic remedies.

It may be administered, with reasonable expectation of success, in the following diseases:

Neuralgia of every description, chronic rheumatism, consumption, scrofula, ophthalmia, chronic ulcerations, and skin diseases, syphilis, and certain affections of the neck of the bladder and prostrate gland, and whenever a powerful alterative agent is needed. This quality of Iodoform is greatly enchanced, in the majority of cases, by the addition of pure iron, Fer. per hydrog.

26

PILLS.	BOTTLE. 100	500
Ipecac. et Opii. 3½ grs. (Pulv. Doveri, U. S. P.) .... Med. prop.—Anodyne, Soporific. Dose, 1 to 3.	50	2 25
Ipecac. et Opii. 5 grs............. Med. prop.—Anodyne, Soporific. Dose 1.	65	3 00
Irisin Comp...................... Med. prop.—Cathartic, Nerve stimulant. Dose, 1 to 2. Irisin, ¼ gr. Podophyllin, 1-10 gr. Strychnine, 1-40 gr.	50	2 25
Laxative........................ Med. prop.—Gentle purgative. Dose, 1 to 2. Pulv. Aloes Soc., 1 gr. Sulphur, 1-5 gr. Res. Podophylli, 2-5 gr. Res. Guaiaci, ½ gr. Syr. Rhamni, q. s.	60	2 75
Leptandrin. ¼ gr................ Med. prop. Cathartic. Dose, 1 to 4.	40	1 75
Leptandrin. ½ gr................ Med. prop.—Cathartic. Dose, 1 to 2.	50	2 25
Leptandrin. 1 gr................ Med. prop.—Cathartic. Dose, 1.	75	3 50
Leptandrin Comp............... Med. prop.—Laxative, Diuretic. Dose, 1 to 2. Leptandrin, 1 gr. Irisin ¼ gr. Podophyllin, ⅛ gr.	1 00	4 75
Lupulin. 3 grs.................. Med. prop.—Anodyne. Dose, 2 to 4.	40	1 75
Morphinæ Acetas. ⅛ gr......... Med. prop.—Anodyne. Dose, 1 to 2.	50	2 25
Morphinæ Comp................. Med. prop. — Anodyne, Febrifuge. Dose, 1. Morph. Sulph., ¼ gr. Ant. et, Pot. Tart., ¼ gr. Hydrarg. Chlor. Mit., ¼ gr.	1 25	6 00
Morphinæ Sulph. 1-20 gr........ Med. prop.—Anodyne. Dose, 1 to 2.	40	1 75
Morphinæ Sulph. 1-10 gr........ Med. prop.—Anodyne. Dose, 1 to 2.	40	1 75

PILLS.	BOTTLE.	
	100	500
Morphinæ Sulph. ⅛ gr.........	50	2 25
Med. prop.—Anodyne. Dose, 1 to 2.		
Morphinæ Sulph. ⅙ gr........	60	2 75
Med. prop.—Anodyne. Dose, 1 to 2.		
Morphinæ Sulph. ¼ gr..........	75	3 50
Med. prop.—Anodyne. Dose, 1 to 2.		
Morphinæ Sulph. ½ gr.........	1 25	6 00
Med. prop.—Anodyne. Dose, 1.		
Morphinæ Valerianas. ⅛ gr...	90	4 25
Med. prop.—Anodyne. Dose, 1 to 2.		
Neuralgic............................	1 90	9 25
Med. prop.—Tonic, Alterative, Ano- dyne. Dose, 1 to 3.		
Quininæ Sulph., 2 grs. Morphinæ Sulph., 1-20 gr, Strychnine, 1-30 gr. Acid Arsenious, 1-20 gr. Ext. Aconiti, ½ gr.		
Neuralgic. (One-half size.).......	1 00	4 75
Med. prop.—As above. Dose, 1 to 3.		
Neuralgic. (Brown Sequard.)......	1 60	7 75
Med. prop.—Anodyne. Dose, 1.		
Ext. Hyoscyami, ⅔ gr. Ext. Conii, ⅔ gr. Ext. Ignat. Amar., ¼ gr. Ext. Opii, ½ gr. Ext. Aconiti, ⅓ gr. Ext. Cannab. Ind., ¼ gr. Ext. Stramon., 1-6 gr. Ext. Belladon., ⅙ gr.		
Neuralgic. (Sine Morphine.)......	1 90	9 25
Med. prop.—Tonic, Alterative. Dose, 1 to 3.		
Opii. ½ gr..........................	40	1 75
Med. prop.—Anodyne. Dose, 1 to 2.		
Opii, U. S. P. 1 gr..............	50	2 25
Med. prop.—Anodyne. Dose, 1.		
Opii et Camphor...............	60	2 75
Med. prop.—Anodyne, Nerve sedative. Dose, 1.		
Pulv. Opii, 1 gr. Camphoræ, 2 grs.		

PILLS.	BOTTLE.	
	100	500
Opii et Camphor et Tannin ... Med. prop.—Anodyne, Astringent. Dose, 1 to 3. Pulv. Opii, ¼ gr. amphoræ, 1 gr. Acid. Tannic, 2 grs.	60	2 75
Opii et Plumbi Acetas .... Med. prop.—Anodyne, Sedative. Dose, 1 to 2. Pulv. Opii, ½ gr. Plumbi Acet., 1½ gr.	50	2 25
Opium, Purified. ½ gr........... Med. prop.—Anodyne, Soporific. Dose, 1 to 2.	1 25	6 00
Phosphori. 1-100 gr., 1-50 gr., 1-25 gr., in each................... Med. prop.—Nerve stimulant. Dose, 1.	1 00	

The method of preparing Phosphorus
in pilular form has been *discovered and
brought to perfection by us*, without the
necessity of combining it with resin, which
forms an insoluble compound. The ele-
ment is in a perfect state of subdivision
and incorporated with the excipient while
in solution. The non-porous coating of
sugar protects it thoroughly from oxidation, so that the pill is not impaired by
age. It is the most pleasant and acceptable form for the administration of Phosphorus.

Phosphori Comp.................. Med. prop.—Nerve tonic. Dose, 1. Phosphori, 1-100 gr. Ext. Nuc. Vom., ¼ gr.	1 25	
Phosphori cum Aloe et Nuc. Vomica,.................... Med. prop.—Useful in the atonic form of Dyspepsia and Neurosis of the Stomach. Dose, 1. Phosphori, 1-50 gr. Ext. Aloe Aq., ¼ gr. Ext. Nuc. Vom., ¼ gr.	1 50	
Phosphori cum Belladonna.. Med. prop.—Useful in Anæmic Condi- tions and Neuralgia. Dose, 1 to 2. Phosphori, 1-100 gr. Ext. Belladonna ⅛ gr.	1 50	

PILLS.	BOTTLE.	
	100	500
Phosphori cum Cannab. Ind	1 75	
Med. prop.—Narcotic, Aphrodisiac.		
Dose, 1 to 2.		
Phosphori, 1-50 gr.		
Ext. Cannab. Ind., ¼ gr.		
Phosphori cum Ferro Comp.	1 50	
Med. prop.—Nutritive, Tonic and Stimulant to the Nervous system. Dose, 1.		
Phosphori, 1-50 gr.		
Strychnine, 1-60 gr.		
Ferri Redact., 1½ gr.		

PHOSPHORUS has recently been prescribed with great advantage in cases of extreme debility and mental depression, from prolonged anxiety or excessive excitement; also, in nervous prostration from overwork, especially brain-work. It is recommended in cases which are attended with great prostration of the vital powers; in exhausting diseases, such as Cholera, Diphtheria, and the latter stages of Typhus and other Fevers, etc. In Epilepsy, Epileptiform-Vertigo, Melancholia, Softening and some other diseases of the Brain, it has been given with marked benefit; also in Neuralgia, Tuberculosis, and Scrofula, in chronic and inveterate diseases of the skin, Leprosy, Lupus, and Psoriasis. Dr. Burgess recommends it in Pruritus Pudendi and other forms of Pruritus.

PHOSPHORUS, STRYCHNINE, AND IRON, combined in the proportions above indicated, is a safe and valuable remedy. As a nutritive tonic and stimulant to the nervous system, especially the spinal cord, it is admirably adapted for the treatment of a large number of nervous disorders dependent on defective nutrition and debility. It increases appetite and promotes digestion. It may be safely given in all those cases in which hypophosphites are employed. It is strongly recommended in *Consumption, Neuralgia, Atonic Dyspepsia, Lowness of Spirit, in General Debility*, and in that general condition of depression and loss of power popularly known as *below par*, and in *breakdown* from overwork and mental fatigue.

PILLS.	BOTTLE.	
	100	500

Phosphori cum Digital. Co ... | 1 50 |
Med. prop.—Valuable as a Heart tonic.
Dose, 1.
Phosphori,   1-50 gr.)
Pv. Digitalis,   1 gr. }
Ext. Hyoscy.,   1 gr.)

Phosphori cum Digitale et Ferro .... | 1 50 |
Med. prop.—Valuable as a Heart tonic.
Dose, 1.
Phosphori,   1-50 gr.)
Pv. Digitalis,   1 gr. }
Ferri Redact.,   1 gr.)

Phosphori cum Ext. Aconiti .. | 1 50 |
Med. prop.—Useful in the Treatment
of Phthisis with Pyrexia. Dose, 1.
Phosphori,   1-50 gr.)
Ext. Aconiti,   1-16 gr. }

Phosphori cum Ferro .... | 1 25 |
Med. prop.—A Powerful Nervine tonic
and Blood restorer. Dose, 1 to 2.
Phosphori,   1-50 gr.)
Ferri Redact.,   1 gr. }

Phosphori cum Cantharide Co | 1 50 |
Med. prop.—Stimulating emmenagogue
and Diuretic. Dose, 1 to 2.
Phosphori,   1-50 gr.)
Pv. Nuc. Vom.,   1 gr. }
Sol. Cantharidis Con.,   1 m.)

Phosphori cum Ferro et Nuc. Vom .... | 1 25 |
Med. prop.—Nerve stimulant, Tonic.
Dose, 1 to 2.
Phosphori,   1-100 gr.)
Ferri Carb.,   1 gr. }
Ext. Nuc. Vom.,   ¼ gr.)

Phosphori cum Ferro et Quinina .... | 1 60 |
Med. prop.—Nerve tonic. Dose, 1 to 2
Phosphori,   1-100 gr.)
Ferri Carb.,   1 gr. }
Quininæ Sul.,   1 gr.)

Phosphori cum Ferro et Quinina et Nuc. Vom .... | 1 60 |
Med. prop.—Nerve tonic. Dose, 1 to 2
Phosphori,   1-100 gr.)
Ferri Carb.,   1 gr. |
Quininæ Sul.,   1 gr. |
Ext. Nuc. Vom.,   ¼ gr.)

PILLS.

	BOTTLE.	
	100	500

Phosphori cum Ferro et Strychnina 1 50
Med. prop.—Nerve tonic and Stimulant.
Dose, 1 to 2.
Phosphori,   1-100 gr. }
Ferri Carb.,   1 gr. }
Strychnin,   1-60 gr. }

Phosphori cum Morphina et Zinco Valer. 2 00
Med. prop.—Nerve tonic and Sedative.
Dose, 1.
Phosphori,   1-50 gr. }
Morphinæ Sul.,   1-12 gr. }
Zinci Valer,   1 gr. }

Phosphori cum Nuc. Vomica. .. 1 25
Med. prop.—Nerve tonic and Stimulant.
Dose, 1 to 2.
Phosphori,   1-50 gr. }
Ext. Nuc., Vom.,   ⅛ gr. }

Phosphori cum Opio et Digital. 1 50
Med. prop.—Useful in Arresting Abnormal calorification. Dose, 1 to 2.
Phosphori,   1-50 gr. }
Pv. Digitalis,   ⅛ gr. }
Pv. Ipecac.,   ¼ gr. }
Pv. Opii,   ¼ gr. }

Phosphori cum Quinina 1 80
Med. prop.—Nerve tonic. Dose, 1 to 2
Phosphori,   1-50 gr. }
Quininæ Sul.,   1 gr. }

Phosphori cum Quininæ Co. .. 1 35
Med. prop.—Nerve tonic. Dose, 1.
Phosphori,   1-50 gr. }
Ferri Redact.   1 gr. }
Quininæ Sul.,   ½ gr. }
Strychnin.,   1-60 gr. }

Phosphori cum Quinina et Ferro et Strychnina 1 60
Med. prop.—Powerful nerve stimulant.
Dose, 1.
Phosphori   1-100 gr. }
Quininæ Sul.,   1 gr. }
Ferri Redacti,   1 gr. }
Strychnin,   1-60 gr. }

PILLS.	BOTTLE.	
	100	500

Phosphori cum Quinina et Digitale Co. .... — 1 35
Med. prop.—Valuable as a Sedative and Diuretic. Dose, 1 to 2.

Phosphori,   1-50 gr. ⎫
Quininæ Sul.,   ½ gr. ⎪
Fv. Digitalis,   ½ gr. ⎬
Pv. Opii,   ¼ gr. ⎪
Pv. Ipecac.,   ¼ gr. ⎭

Phosphori cum Quinina et Nuc Vom .... — 1 60
Med. prop.—Nerve tonic. Dose, 1 to 2

Phosphori,   1-50 gr. ⎫
Quininæ Sul.,   1 gr. ⎬
Ext. Nuc. Vom.,   ¼ gr. ⎭

Phosphori cum Strychnina ... — 1 25
Med. prop.—Nerve tonic and Stimulant. Dose, 1.

Phosphori,   1-50 gr. ⎫
Strychnin,   1-60 gr. ⎭

Phosphori cum Zinco Co. ...... — 1 50
Med. prop.—Useful in Uterine disturbances, Leucorrhœa and Hysteria. Dose, 1 to 2.

Phosphori,   1-50 gr. ⎫
Zinci Sul.,   1 gr. ⎬
Lupulin,   1 gr. ⎭

Phosphori et Damiana cum Nuc. Vom .... — 1 50
Med. prop.—Aphrodisiac. Dose, 1 to 2.

Ext. Damiana,   2 grs. ⎫
Phosphori,   1-100 gr. ⎬
Ext. Nuc. Vom.,   ⅛ gr. ⎭

Podophyllin. 1-10 gr .... — 40 | 1 75
Med. prop.—Cathartic. Dose, 1 to 4.

Podophyllin. ¼ gr. .... — 40 | 1 75
Med. prop.—Cathartic. Dose, 1 to 4.

Podophyllin. ½ gr .... — 50 | 2 25
Med. prop.—Cathartic. Dose, 1 to 2.

Podophyllin. 1 gr .... — 75 | 3 50
Med. prop.—Cathartic. Dose, 1.

33

PILLS.	BOTTLE.	
	100	500
Podophylli. (Dr. E. R. Squibb.)...	75	3 50

Dose, 1 to 2.
Res. Podophylli, ¼ gr.
Pv. Capsici, ½ gr.
Ext. Belladon., ⅓ gr.
Pv. Sacch. Lact., 1 gr.
Acacia and Glycerin., aa q. s.

The composition of this pill is the same as Dr. Squibb's recipe, and for administration is preferable, being just as soluble and more elegant in appearance.

Practical test for Solubility:—Suspend 3 of these pills in water and they will dissolve, coating included, as soon as the plain pills ready prepared. This is accomplished through our method of manipulation and coating, and physicians should not regard the prejudice against sugar-coated pills as applying to Wm. R. Warner & Co.'s manufacture.

Podophyllin et Bellad..........	75	3 50

Med. prop.—Mild stimulating laxative.
Dose, 1 to 3.
Podophyllin, ¼ gr.
Ext. Bellad., ⅔ gr.
Ol. Res. Capsici, ¼ gr.
Saccharum Lact., 1 gr.

Podophyllin Comp............	75	3 50

Med. prop.—Cathartic and Tonic.
Dose, 1 to 2.
Podophyllin, ¼ gr.
Ext. Hyoscyami, ⅓ gr.
Ext. Nuc. Vomicæ, 1-16 gr.

Podophyllin Comp. (Eclectic.)..	75	3 50

Med. prop.—Purgative. Dose, 2 to 4.
Podophyllin, ⅓ gr.
Leptandrin, 1-16 gr.
Juglandin, 1-28 gr.
Macrotin, 1-32 gr.
Ol. Res. Capsici, q. s.

Podophyl. et Hydrarg.........	50	2 25

Med. prop.—Laxative. Dose, 2 to 4.
Podophyllin, ¼ gr.
Pil. Hydrarg., 2 grs.

Podophyllin et Hyoscyamus.	60	2 75

Med. prop.—Gentle cathartic. Dose, 1 to 2.
Podophyllin, ½ gr.
Ext. Hyoscyami, ½ gr.

PILLS.	BOTTLE.	
	100	500
Podophyllotoxin. ⅛ gr.......... Med. prop.—Cathartic, without the nausea induced by Podophyllin. Dose, 1 to 2.	50	2 25
Podophyllotoxin. ¼ gr.......... Med. prop.—Cathartic, without the nausea induced by Podophyllin. Dose, 1 to 2.	75	3 50
Post-Partum. (Dr. Fordyce Barker.) Dose, 1. Ext. Coloc. Comp., 1⅛ gr. Hydrarg. Chlor. Mit., 1½ gr. Ext. Hyoscyami, ⅓ gr. Ext. Nuc. Vom., ⅛ gr. Pulv. Aloes, ⅙ gr. Pulv. Ipecac,. ⅙ gr.	1 00	4 75
Potass. Bromid. 1 gr..... ...... Med. prop.—Nerve sedative. Dose, 2 to 5.	75	3 50
Potass. Bromid. 5 grs............ Med. prop.—Nerve sedative. Dose, 1 to 2.	1 25	6 00
Potass. Iodid. (Merck's) 2 grs.... Med. prop.—Alterative. Dose, 1 to 3.	85	4 00
Potass. Permanganas. ⅛ gr... Med. prop.—Dr. ROBERTS BARTHOLOW, Professor of Materia Medica and Therapeutics in the Jefferson Medical College of Philadelphia, says ("Permanganate of Potassium, its Action and Uses"):— "One of the most important therapeutical applications of permanganate of potassium, a recent discovery, is in the treatment of amenorrhœa. We owe this valuable improvement, as indeed many others, to Drs. Ringer and Murrell. They have shown that this remedy is remarkably certain when applied in suitable cases. Given in doses of two to five grains three times a day, for several days preceding the menstrual molimen, this *agent is quite sure to start the flow.* Dose, 2 to 5 grs.	50	2 25
Potass. Permanganas. 1 gr.... Med. prop.—As above. Dose, 1 to 4.	75	3 50
Potass. Permanganas. 2 grs.... Med. prop.—As above. Dose, 1 to 2.	1 00	4 75

PILLS.

	BOTTLE.	
	100	500
Prandii............................	75	3 50
Med. prop.—Stimulating purgative.		
Dose, 1 to 2.		
Ext. Aloe Aq., 1 gr.		
Ext. Gentian, 2 grs.		
Ext. Anthemid., 1 gr.		
Pv. Capsici, ¼ gr.		
Quininæ Bi-Sulph. ½ gr.........	35	1 50
Med. prop.—Tonic, Anti-periodic.		
Dose, 1 to 4.		
Quininæ Bi-Sulph. 1 gr.........	40	1 75
Med. prop.—Tonic, Anti-periodic.		
Dose, 1 to 3.		
Quininæ Bi-Sulph. 2 grs........	60	2 75
Med. prop.—Tonic, Anti-periodic.		
Dose, 1 to 3.		
Quininæ Bi-Sulph. 3 grs........	85	4 00
Med. prop.—Tonic, Anti-periodic.		
Dose, 1 to 2.		
Quininæ Bi-Sulph. 5 grs........	1 50	
Med. prop.—Tonic, Anti-periodic.		
Dose, 1 to 2.		
Quininæ Comp...................	1 00	4 75
Med. prop.—Tonic, Anti-periodic.		
Dose, 1 to 2.		
Quininæ Sulph., 1 gr.		
Fer. Carb. (Vallett's), 2 grs.		
Acid Arsenious, 1-60 gr.		
Quininæ cum Capsicum........	80	3 75
Med. prop.—Anti-periodic, Stimulant.		
Dose, 1 to 3.		
Quininæ Sulph., 1 gr.		
Capsicum, ¼ gr.		
Quininæ et Ext. Bellad........	1 00	4 75
Med. prop.—Nerve tonic, Anti-periodic.		
Dose, 1 to 2.		
Quininæ Sulph., 1 gr.		
Ext. Belladon., ½ gr.		
Quininæ et Ferri..............	80	3 75
Med. prop.—Tonic, Anti-periodic.		
Dose, 1 to 2.		
Quininæ Sulph., 1 gr.		
Ferrum per Hydrog., 1 gr.		
Quininæ et Ferri Carb........	1 00	4 75
Med. prop.—Tonic, Anti-periodic.		
Dose, 1 to 2.		
Quininæ Sulph., 1 gr.		
Fer. Carb. (Vallett's), 2 grs.		

36

PILLS.	BOTTLE.	
	100	500
Quininæ et Ferri Cit. 1 gr..... Med. prop.,—Tonic, Anti-periodic. Dose, 1 to 2.	40	1 75
Quininæ et Ferri Cit. 2 grs.... Med. prop.—Tonic, Anti-periodic. Dose, 1 to 2.	50	2 25
Quininæ et Fer. et Strychnin..................... Med. prop.,—Tonic, Anti-periodic. Dose, 1 to 2. Quininæ Sulph., 1 gr. } Fer. Carb. (Vallett's), 2 grs. } Strych. Sulph., 1-60 gr. }	80	3 75
Quininæ et Fer. et Strych. Phos...................... Med. prop.—Tonic, Anti-periodic. Dose, 1 to 2. Quininæ Phos., 1 gr. } Ferri Phos., 1 gr. } Strychninæ Phos., 1-60 gr. }	1 00	4 75
Quininæ et Ferri Valer. 2 gr... Med. prop.—Tonic, Nerve sedative. Dose, 1 to 2.	1 75	8 50
Quininæ et Hydrarg............ Med. prop.—Tonic, Anti-periodic. Dose, 1 to 2. Quininæ Sulph., 1 gr. } Mass. Hydrarg., 2 grs. } Oleo-Res. Piper. Nig., ¼ gr. }	1 00	4 75
Quininæ et Iodoform .... Med. prop.—Tonic, Alterative. Dose, 1 to 3. Iodoform, ½ gr. } Quininæ Bi-Sulph., ½ gr. }	1 50	
Quininæ Iodoform et Fer..... Med. prop.—Tonic, Alterative. Dose, 1 to 2. Iodoform, 1 gr. } Fer. Carb. (Vallett's), 2 grs. } Quininæ Sulph., ½ gr. }	1 25	6 00
Quininæ et Strychnin......... Med. prop.—Tonic, Nerve stimulant. Dose, 1 to 2. Quininæ Sulph., 1 gr. } Strychninæ, 1-60 gr. }	80	3 75

PILLS.	BOTTLE.	
	100	500

Quininæ et Strychnin Comp.. 1 00 | 4 75
Med. prop.—Tonic, Alterative. Dose, 1.

Quininæ Sulph.,	1 gr.
Ferri per Hydrog.,	1½ gr.
Strychnine,	1-20 gr.
Acid Arsenious,	1-20 gr.

Quininæ Sulph. ½ gr............ 35 | 1 50
Med. prop.—Tonic, Anti-periodic.
Dose, 1 to 4.
Our Quinine Pills are manufactured from
Powers & Weightman's Quinine, and other
brands having an established reputation.
An examination by a competent Chemist
is made to verify our tests. Each Pill
contained in packages to which our label
is attached, is warranted to have the full
complement of pure material as expressed
thereon.

Quininæ Sulph. 1 gr............ 40 | 1 75
Med. prop.—Tonic, Anti-periodic.
Dose, 1 to 3.

Quininæ Sulph. 2 grs............ 60 | 2 75
Med. prop.—Tonic, Anti-periodic.
Dose, 1 to 3.

Quininæ Sulph. 3 grs............ 85 | 4 00
Med. prop.—Tonic, Anti-periodic.
Dose, 1 to 2.

Quininæ Sulph. 5 grs............ 1 50
Med. prop.—Tonic, Anti-periodic.
Dose, 1 to 2.

Quininæ Valerianas. ½ gr..... 1 45 | 7 00
Med. prop.—Tonic, Nervine Dose, 1
to 2.

Quinidinæ Sulph. 1 gr.......... 1 15 | 5 50
Med. prop.—Anti-periodic. Dose, 1 to 3

Quinidinæ Sulph. 2 grs........ 2 15 | 10 50
Med. prop.—Anti-periodic. Dose, 1 to 3

Quinidinæ Sulph. 3 grs........ 3 25 | 16 00
Med. prop.—Anti-periodic. Dose, 1 to 2

Rhei Comp., U. S. P............. 75 | 3 50
Med. prop.—Purgative. Dose, 2 to 4.

Pulv. Rhei,	2 grs.
Pur. Aloes,	1½ gr.
Myrrh.,	1 gr.
Ol. Menth. Pip.,	1-10 gr.

PILLS.

	BOTTLE.	
	100	500
Rhei et Hydrarg............. Med. prop.—Cholagogue cathartic. Dose, 2 to 5. Pulv. Rhei. Mass. Hydrarg. Sodii Carb. Exsic.	80	3 75
Rhei, U. S. P.................... Med. prop.—Gentle laxative. Dose, 2 to 5. Pulv. Rhei, 3 grs. } Pulv. Saponis, 1 gr. }	75	3 50
Rheumatic......................... Med. prop.—Anti-rheumatic, Purgative. Dose, 1 to 3. Ext. Coloc. Comp., 1½ gr. } Ext. Colchici Acet., 1 gr. } Ext. Hyoscyami, ⅓ gr. } Hydrarg. Chlor. Mit., ⅓ gr. }	90	4 25
Salicylic Acid. 2½ grs............ Med. prop.—Anti-rheumatic. Dose, 1 to 2.	75	3 50
Salicylic Acid. 5 grs............ Med. prop.—Anti-rheumatic. Dose, 1 to 2.	1 30	6 25
Santonin. 1 gr.................... Med. prop.—Anthelmintic. Dose, 1 to 3.	1 00	4 75
Scillæ Comp., U. S. P.......... Med. prop.—Expectorant, Diuretic. Dose, 1 to 3. Pulv. Scillæ, ½ gr. } Pulv. Zingiber. Jam., 1 gr. } Gum Ammoniac, 1 gr. } Pulv. Saponis, 1½ gr. }	50	2 25
Sedative......................... Med. prop.—Sedative. Dose, 1 to 2. Ext. Sumbul, ½ gr. } Ext. Valerian, ½ gr. } Ext. Hyoscyami, ½ gr. } Ext. Cannab. Ind., 1-10 gr. }	75	3 50
Silver Nitras. ¼ gr........... Med. prop.—Alterative to Mucous Membrane. Dose, 1 to 4.	75	3 50
Silver Iodid, ¼ gr............ Med. prop.—Alterative to Mucous Membrane. Dose, 1 to 4.	75	3 50

39

PILLS.	BOTTLE.	
	100	500
Stomachicw. (Lady Webster.)	50	2 25
Med. prop.—Stimulating purgative.		
Dose, 1 to 2.		
Aloe Socot., 2 grs.)		
Gum Mastich., ½ gr. }		
Fol. Rosæ, ½ gr.)		
Strychnina. 1-60 gr. 1-40 gr. 1-32 gr.	40	1 75
1-30 gr. 1-20 gr. 1-16 gr.		
Med. prop.—Nerve stimulant, Tonic.		
Dose, 1 to 3.		
Strychnine Sulph. 1-32 gr.	40	1 75
Med. prop.—Nerve stimulant, Tonic.		
Dose, 1 to 3.		
Syphilitic	1 00	4 75
Med. prop.—Specific alterative. Dose,		
1 to 2.		
Potass. Iodid., 2½ grs.)		
Hyd. Chlor. Cor., 1-40 gr.)		
Tonic	60	2 75
Med. prop.—Tonic. Dose, 2 to 3.		
Ext. Gentian, 1 gr.)		
Ext. Humuli, ½ gr.		
Ferri Carb. Sacch., ¼ gr.		
Ext. Nuc. Vom., 1-20 gr (		
Res. Podophylli, 1-25 gr.		
Ol. Res. Zingiber., 1-10 gtt.)		
Triplex	75	3 50
Med. prop.—Purgative. Dose, 2 to 4.		
Aloe Socot., 2 grs.)		
Mass. Hydrarg., 1 gr. }		
Podophyllin, ¼ gr.)		
Triplex Improved	75	3 50
Med. prop.—Purgative. Dose, 2 to 4.		
Pv. Scammon. Virg., 1 1-5 gr.)		
Pv. Aloes Soct., 1 1-5 gr.		
Pil. Hydrarg., 1 1-5 gr.		
Ol. Tiglii, 1-20 m.		
Ol. Carui, ¼ m.		
Tinct. Aloes et Myrrh., q. s.)		
Zinc Phosphide. ⅙, ¼ gr.	75	3 50
Med. prop.—Tonic. Dose, 1 to 3.		
Zinci Phosphide et Nuc. Vom.	1 00	4 75
Med. prop.—Tonic, Stimulant. Dose,		
1 to 3.		
Zinci Phos., 1-16 gr.)		
Ext. Nuc. Vom., ¼ gr.)		
Zinci Valerianas. 1 gr.	1 00	4 75
Med. prop.—Anti-spasmodic. Dose, 1		
to 3. 40		

PILLS.	BOTTLE	
	100	500
Aloes, U. S. P. .... Med. prop.—Stimulating, Purgative. Dose, 1 to 3.	40	1 75
Aloes et Asafoetida, U. S. P. ... Dose, 2 to 5.	40	1 75
Aloes et Ferri, U. S. P. .... Med. prop.—Tonic, Purgative. Dose, 1 to 3. Pur. Aloes, 1 gr. Aromat Powd. 1 gr. Ferri Sulph. Exs. 1 gr. Confect. Rose, q. s.	40	1 75
Aloes et Mastich. (Lady Webster) Med. prop.—Stimulating purgative. Dose, 1 to 2.	50	2 25
Aloes et Myrrhæ, U. S. P ..... Med. prop.—Cathartic, Emmenagogue. Dose, 3 to 6.	50	2 25
Aloes et Nuc. Vomicæ .... Med. prop.—Tonic, Purgative. Dose, 1 to 2. Pulv. Aloes Soc. 1½ grs. Ext. Nuc. Vom. ¼ gr.	50	2 25
Aloïn et Strychnin. et Belladon .... Med. prop.—Tonic, Laxative. Aloin, ¼ gr. Strychnine, 1-60 gr. Ext. Belladon. ⅛ gr. Used very largely and with great success in the treatment of habitual constipation.	60	2 75
Alterative .... Med. prop.—Alterative with tendency to mercurial impression. Dose, 1 to 2. Mass. Hydrarg. 1 gr. Pulv. Opii, ⅛ gr. Pulv. Ipecac, ⅛ gr.	50	2 25
Anti-Bilious. (Vegetable) .... Med. prop.—Cholagogue, Cathartic. Dose, 2 to 3. Pulv. Ext. Coloc. Co. 2½ grs. Podophyllin, ¼ gr.	50	2 25

41

PILLS.	BOTTLE	
	100	500
Antiperiodic......................	80	3 75
Med. prop.—Antiperiodic. Dose, 1 to 2		
Cinchona Sul. 1 gr.		
Podophyl. 1-20 gr.		
Strych. Sul. 1-33 gr.		
Gelsemin. 1-20 gr.		
Ferri Sul. Ext. ½ gr.		
Ol. Res. Caps. 1-10 gtt.		
Aperient.........................	85	4 00
Med. prop.—Aperient, Tonic. Dose,		
1 to 2.		
Ext. Nuc. Vom. ⅓ gr.		
Ext. Hyoscyami ½ gr.		
Ext. Coloc. Comp, 2 grs.		
Aperient, (Dr. Fordyce Barker)....	1 00	4 75
Med. prop.—Aperient.		
Ext. Coloc. Comp. 1¾ grs.		
Ext. Nuc. Vom. ½ gr.		
Ext. Hyoscyam. 1¼ grs.		
Pulv. Ipecac. 1-12 gr.		
Pulv. Aloes Soc. 5-12 gr.		
Res. Podophylli, 1.12 gr.		
Asafoetida, U. S. P..............	40	1 75
Med. prop.—Nerve stimulant. Dose,		
1 to 3.		
Camphor et Ext. Hyoscyamus	50	2 25
Med. prop.—Anodyne, Cerebral stimu-		
lant. Dose, 1 to 2.		
Camphor, 1 gr.		
Ext. Hyoscyami, Eng. 1 gr.		
Cascara Comp...................	75	3 50
Med. prop.—Laxative, Cathartic.		
Dose, 2 to 4.		
Ext. Cascara Sagrad. 3 grs.		
Res. Podophylli, ⅛ gr.		
Cathart. Comp. U. S. P.........	30	1 35
Med. prop.—Cathartic. Dose, 2 to 4.		
These pills are made strictly in accord-		
ance with the formula as directed by the		
Pharmacopœia.		
Cathart. Comp. Improved.....	30	1 35
Med. prop.—Cathartic. Dose, 2 to 4.		
Ext. Coloc. Comp.		
Ext. Jalap.		
Podophyllin, Leptandrin.		
Ext. Hyoscyami.		
Ext. Gentian.		
Ol. Menth. Pip. 3 grs.		

PILLS.	BOTTLE	
	100	500
Cathart. Comp. Vegetable	30	1 35
Med. prop.—Cathartic. Dose, 2 to 3.		
Podophyllin, Scammony,)		
Ext. Colocynth, }		
Aloes, Soap and Card. 3 grs.)		
Cinchoniæ Sulph. 1¼ grs.........	60	2 75
Med. prop.—Tonic, Antiperiodic.		
Dose, 1 to 3.		
Cinchonidiæ Salicylate, 2½ grs.	1 50	7 25
Med. prop.—Antirheumatic. Dose, 1		
to 2.		
Cinchonidiæ Sulph. 1 gr.........	40	1 75
Med. prop.–Antimalarial, Antiperiodic.		
Dose, 1 to 3.		
Cinchonidiæ Sulph. 2 grs.......	45	2 00
Med. prop.-Antimalarial, Antiperiodic.		
Dose, 1 to 3.		
Cinchonidiæ Sulph. 3 grs.......	50	2 25
Med. prop.-Antimalarial, Antiperiodic.		
Dose, 1 to 2.		
Colocynth et Hydrarg.		
et Ipecac	75	3 50
Med. prop.—Cholagogue, Cathartic.		
Dose, 1 to 3.		
Pulv. Ext. Coloc. Co. 2 grs.)		
Pil. Hydrarg. 2 grs. >		
Pulv. Ipecac, ⅛ gr.)		
Colocynth et Hyoscyamus	75	3 50
Med. prop.—Gentle laxative. Dose,		
1 to 2.		
Ext. Coloc. Comp. 2½ grs.)		
Ext. Hyoscyami, 1⅛ grs.)		
Damianæ cum Phos.		
et Nuc Vom	1 50	7 25
Med. prop.—Aphrodisiac. Dose, 1 to 2.		
Ext. Damianæ, 2 grs.)		
Phosphori, 1-100 gr. -		
Ext. Nuc. Vom. ⅛ gr.)		
A valuable remedy, indicated in sexual		
debility, overwork of the brain, impo-		
tency, etc. It is also highly recommended		
as a uterine tonic. Also of value in Leu-		
corrhœa, Amenorrhœa, Dysmenorrhœa,		

PILLS.	BOTTLE	
	100	500

Digestiva..................................... 75 | 3 50
Med. prop.—Useful in Indigestion.
Dose, 1 to 2.
Pepsin Concentrat. 1 gr. ⎫
Pulv. Nuc. Vom. ¼ gr. ⎪
Gingerine, 1-10 gr. ⎬
Sulphur, ⅓ gr. ⎭
This combination is very useful in relieving various forms of Dyspepsia and Indigestion, and will afford permanent benefit in cases of enfeebled digestion, where the gastric juices are not properly secreted.

As a corrective of nausea or lack of appetite in the morning, induced by over-indulgence in food or stimulants during the night, these pills are unsurpassed. They should be taken in doses of two pills before retiring, or in the morning at least one hour before eating; the first mentioned time is the most desirable, as the effects are more decided, owing to the longer period for action, and the natural rest is more fully experienced through their mild but effective influence.

As a dinner pill, Pil. Digestiva is unequalled, and may be taken in doses of a single pill either before or after eating.

Digitalin (Alkaloid) 1-60 gr......... 75 | 3 50
Med. prop.—Arterial sedative. Dose, 1 to 2.

Emmenagogue.................... 1 25 | 6 00
Med. prop.—Active emmenagogue. Tonic. Dose, 1 to 3.
Ergotine, 1 gr. ⎫
Ext. Hellebor. Nig. 1 gr. ⎪
Aloes, 1 gr. ⎬
Ferri Sulph. Exs. 1 gr. ⎪
Ol. Sabinæ, ½ gr. ⎭

Ergotin, 1 gr.......................... 1 00 | 4 75
Dose, 1 to 4.

Ergotin, 3 grs......................... 1 50 | 7 25
Med. prop.—Parturient. Dose, 1 to 2.

Ext. Colocynth Comp. 3 grs.... 80 | 3 75
Dose, 1 to 3.

Ferrum (Quevenne's) 2 grs.......... 75 | 3 50
Med. prop.—Tonic. Dose, 1 to 2.

44

PILLS.	BOTTLE	
	100	500
Gonorrhoea............................	60	2 75
Med. prop.—Tonic, Alterative to Mucous Membrane. Dose, 1 to 3.		
Pulv. Cubeb. 2 grs. Bals. Copaib. Solid, 1 gr. Ferri Sulph. ½ gr. Terebinth Venet. 1½ grs.		
Hepatica..............................	80	3 75
Med. prop.—Cholagogue, Cathartic. Dose, 1 to 2.		
Pil. Hydrarg. 3 grs. Ext. Coloc. Comp. 1 gr. Ext. Hyoscyami, 1 gr.		
Hydrargyri, 5 grs..................	50	2 25
Dose, 1 to 2.		
Hydrargyri. U. S. P. 3 grs......	40	1 75
Med. prop.—Mercurial alterative. Dose, 2 to 3.		
Ipecac et Opii, (Pulv. Doveri, U. S. P.) 3½ grs................	50	2 25
Med. prop.—Anodyne, Soporific. Dose, 1 to 3.		
Ipecac et Opii. 5 grs.............	65	3 00
Med. prop.—Anodyne, Soporific. Dose, 1 to 2.		
Laxative............................	60	2 75
Med. prop.—Gentle purgative. Dose, 1 to 2.		
Pulv. Aloes Soc. 1 gr. Sulphur, 1-5 gr. Res. Podophylli, 1-5 gr. Res. Guaiaci, ½ gr. Syr. Rhamni, q. s.		
Morphiæ Sulph., 1-20 gr.......	40	1 75
Med. prop.—Anodyne. Dose, 1 to 2.		
Morphiæ Sulph., 1-10 gr........	40	1 75
Med. prop.—Anodyne. Dose, 1 to 2.		
Morphiæ Sulph., ⅛ gr...........	50	2 25
Med. prop.—Anodyne. Dose, 1 to 2.		
Morphiæ Sulph., ⅙ gr...........	60	2 75
Med. prop.—Anodyne. Dose, 1 to 2.		
Morphiæ Sulph., ¼ gr...........	75	3 50
Med. prop.—Anodyne. Dose, 1		

45

PILLS.	BOTTLE 100	500
Morphiæ Sulph.. ½ gr.....,.....	1 25	6 00
Med. prop.—Anodyne. Dose, 1 to 2.		
Morphiæ Valerianas. ⅛ gr......	90	4 25
Med. prop.—Anodyne. Dose, 1 to 2.		
Neuralgic, (one-half size)...........	2 00	4 75
Med. prop.—Tonic, Alterative, Anodyne. Dose, 1 to 3		
Neuralgic	1 90	9 25
Med. prop.—Tonic, Alterative, Anodyne. Dose, 1 to 3. Quiniæ Sulph. 2 grs. Morphiæ Sulph. 1-20 gr. Strychnine, 1-30 gr. Acid Arsenious, 1-20 gr. Ext. Aconiti, ½ gr.		
Neuralgic, (Brown-Sequard)........	1 60	7 75
Med. prop.—Anodyne. Dose, 1. Ext. Hyoscyami, ⅔ gr. Ext. Conii, ⅔ gr. Exs. Ignat. Amar. ¼ gr Ext. Opii, ½ gr. Ext. Aconiti, ⅓ gr. Ext. Cannab. Ind. ¼ gr. Ext. Strammon. 1-5 gr Ext. Belladon. ⅙ gr.		
Opii, U.S. P., 1 gr.	50	2 25
Medp. pro.—Anodyne. Dose, 1.		
Opii et Camphor.	60	2 75
Med. prop.—Anodyne, Nerve sedative. Dose, 1. Pulv. Opii, 1 gr. Camphoræ, 2 grs		
Pepsin et Bismuth.............	1 25	6 00
Dose, 1 to 2. Pepsin Sacch, 2 grs. Bis. Sub. Nit., 3 grs.		
Pepsin Sacch	1 00	4 75
Each containing 3 grs. Dose, 1 to 2.		
Pepsin Sacch...................	1 00	4 75
Each containing 5 grs. Dose, 1.		
Phosphori Comp...............	1 25	
Med. Pro.—Nerve tonic. Dose, 1. Phosphori, 1-100 gr. Ext. Nuc. Vom. ¼ gr.		

46

PILLS.	BOTTLE	
	100	500

Phosphori cum Cantharide Comp 1 50
Med. prop.—Stimulating emmenagoge and diuretic.
Phosphori,   1-50 gr. ⎫
Pulv. Nuc. Vom.   1 gr. ⎬
Sol. Cantharidis.,   1 m. ⎭

Phosphori cum Ferro 1 25
Med. prop.—A powerful nervine tonic, and blood restorer. Dose, 1 to 2.
Phosphori,   1-50 gr. ⎫
Ferri Redact,   1 gr. ⎭

Phosphori cum Ferro et Nuc. Vom. 1 25
Med. prop.—Nerve stimulant, Tonic. Dose, 1 to 2.
Phosphori,   1-100 gr. ⎫
Ferri Carb,   1 gr. ⎬
Ext. Nuc. Vom.   ¼ gr. ⎭

Phosphori cum Ferro et Quinina. 1 60
Med. prop.—Nerve Tonic. Dose, 1 to 2
Phosphori,   1-100 gr. ⎫
Ferri Carb.,   1 gr. ⎬
Quinine Sulph.,   1 gr. ⎭

Phosphori cum Ferro et Quinina et Nuc. Vom. 1 60
Med. prop.—Nerve Tonic, Dose, 1 to 2.
Phosphori,   1-100 gr. ⎫
Ferri Carb.,   1 gr. ⎬
Quinæ Sulph.,   1 gr. ⎬
Ext. Nuc. Vom.   ¼ gr. ⎭

Phosphori cum Quinina et Ferro et Strychnina 1 60
Med. prop.—Powerful Nerve stimulant. Dose, 1.
Phosphori,   1-100 gr. ⎫
Quiniæ, Sulph.,   1 gr. ⎬
Ferri Redacti,   1 gr. ⎬
Strychnina,   1-60 gr. ⎭

Phosphori cum Ferro et Strychnia 1 50
Med. prop.—Nerve tonic and stimulant. Dose, 1 to 2.
Phosphori,   1-100 gr. ⎫
Ferri Carb.,   1 gr. ⎬
Strychnina.,   1-60 gr. ⎭

PILLS.	BOTTLE	
	100	500

Phosphori cum Nuc. Vom......   1 35
Med. prop.—Nerve tonic and stimu-
   lant. Dose, 1.
   Phosphori,   1-50 gr. }
   Ext. Nuc. Vom,   ⅛ gr. }

Phosphori cum Quiniæ
     Comp. 1 35
Med. prop.—Nerve tonic Dose, 1.
   Phosphori,   1-50 gr. }
   Ferri Redact.,   1 gr. }
   Quiniæ Sulph.,   ¼ gr. }
   Strychnia,   1-60 gr. }

Podophylli (Warner & Co.)........   75   1 50
Dose, 1 to 2.
   Res Podophylli,   ¼ gr. }
   Pulv. Capsici,   ½ gr. }
   Ext. Belladona.,   ⅛ gr. }
   Pulv. Sacch Lact.,   1 gr. }
   Acacia and Glycerine, aa q. s. }
   The composition of this pill is the same
as Dr. Squibb's recipe, and for adminis-
tration it is preferable, being just as
soluble and more elegant in appearance.
   PRACTICAL TEST FOR SOLUBILITY.
Suspend three of these pills in water, and
they will dissolve, coating included as soon
as the plain pills ready prepared. This
is accomplished through our method of
manipulation and coating, and physicians
should not regard the prejudice against
sugar-coated pills as applying to Wm. R.
Warner & Co.'s manufacture.

Quiniæ Bi-Sulph., ½ gr.........   35   1 50
Med. prop.—Tonic, Anti-periodic
   Dose, 1 to 4.

Quiniæ Bi-Sulph., 1 gr..........   40   1 75
Med. prop.—Tonic, Anti-periodic.
   Dose, 1 to 3.

Quiniæ Bi-Sulph., 2 grs   60   2 75
Med. prop.—Tonic, Anti-periodic.
   Dose, 1 to 3.

Quiniæ Bi-Sulph., 3 grs.........   85   4 00
Med. prop.—,Tonic Anti-periodic.
   Dose, 1 to 2.

PILLS.	BOTTLE	
	100	500

Quiniæ Bi-Sulph., 5 grs.......... 1 50
Med. prop.—Tonic Anti-periodic.
Dose, 1 to 3.

Quiniæ cum Capsicum.......... 80 3 75
Med. prop.—Anti-periodic, Stimulant.
Dose, 1 to 3.
Quiniæ Sulph.,   1 gr. }
Capsicum,   ¼ gr. }

Quinæ et Ferri et Strychnin 80 3 75
Med. prop.—Tonic Anti- periodic.
Dose, 1 to 2.
Quiniæ,   1 gr. }
Ferri Carb. (Vallett's),   2 gr. }
Strych. Sulph.,   1-60 gr. }

Quiniæ et Ferri et Strychnin
Phos. 1 00 4 75
Dose, 1 to 2.
Quiniæ Phos.,   1 gr. }
Ferri Phos.,   1 gr. }
Strych. Phos.,   1-60 gr. }

Quiniæ Sulph., ½ gr............. 35 1 50
Med. prop.—Tonic, Anti-periodic.
Dose, 1 to 4.
Our Quinine Pills are manufactured
from Powers & Weightman's Quinine
and other brands having an established
reputation. An examination by a competent chemist is made to verify our
tests. Each pill, contained in package,
to which our label is attached, is warranted to have the full complement of
pure material as expressed thereon.

Quiniæ Sulph., 1 gr.............. 40 1 75
Med. prop.—Tonic, Anti-periodic.
Dose, 1 to 3.

Quiniæ Sulph., 2 grs. 60 2 75
Med. prop.—Tonic, Anti-periodic.
Dose, 1 to 3.

Quiniæ Sulph., 3 grs. 85 4 00
Med. prop.—Tonic, Anti-periodic.
Dose, 1 to 2.

Quiniæ Sulph. 5 grs. 1 50
Med pro.—Tonic, Anti-periodic.
Dose, 1 to 2.

PILLS.	BOTTLE	
	100	500

Salicylic Acid, 2½ grs..... .... 75 3 50
Med. prop.—Anti-rheumatic.
   Dose, 1 to 2.

Salicylic Acid, 5 grs............... 1 30 6 25
Med. prop.— Anti-rheumatic.
   Dose, 1.

Salol, 2½ and 5 grs................. 1 20 5 75
   Dose, 1 to 2.

Salutis (Dr. T. G. Thomas) 80 3 75
   Dose, 1.
   Aloin,   ½ gr.
   Ext. Hyoscyam,   ½ gr.
   Ext. Nuc· Vom.   ⅛ gr.
   Ext. Belladon,   1-16 gr.

Sedative 75 3 50
Med. prop.—Sedative.
   Ext. Sumbul,   ½ gr
   Ext. Valerian,   ½ gr.
   Ext. Hyoscyami,   ½ gr.
   Ext. Cannab. Ind., 1-10 gr.

Strychnia. 1-20, 1-40, 1-60 gr. 40 1 75
   Dose, 1 to 2.

Tonic (Dr. Aiken) 1 25 6 00
Med. prop.—Tonic.
   Dose, 1.
   Quinin Sul.,   1 gr.
   Ac Arsenious,   1-50 gr.
   Strychnine,   1-50 gr.
   Ferri Redacti,   ⅔ gr.

Tonic—Laxative (Dr. Skene)...... 80 3 75
   Dose, 1.
   Quininæ Sul.   1 gr.
   Ext. Colocynth,   ½ gr.
   Ext Belladon.,   1-10 gr.

WM. R. WARNER & CO.

MANUFACTURING CHEMISTS,

PHILADELPHIA AND NEW YORK.

WARNER & CO.'S

SOLUBLE COATED

GRANULES.

GRANULES.	BOTTLE.	
	100	500
Acid Arsenious. 1-20, 1-30, 1-50, and 1-60 gr........................ Medical properties.—Anti-periodic, Alterative. Dose, 1 to 2.	40	1 75
Aconitia. 1-60 gr........................ Med. prop.—Nerve sedative. Dose, 1 to 2.	75	3 50
Aloin et Strychnin................. Med. prop.—Tonic laxative. Dose, 1 to 2.	60	2 75
Aloin et Strychnin. et Belladon........................ Med. prop.—Tonic, Laxative. Dose, 1 to 2. Aloin, 1-5 gr. Strychnine, 1-60 gr. Ext. Belladon., ¼ gr.	60	2 75
Atropina. 1-60 gr........................ Med. prop.—Anodyne. Dose, 1 to 2.	75	3 50
Atropina. 1-100 gr........................ Med. prop.—Anodyne. Dose, 1 to 2.	75	3 50
Atropinæ Sulph. 1-60 gr.......... Med. prop.—Same as Atropina. Dose, 1 to 2.	75	3 50
Caulophyllin. 1-10 gr............. Med. prop.—Emmenagogue. Dose, 1 to 4.	40	1 75
Cimicifugin. 1-10 gr.............. Med. prop.—Tonic, Nerve stimulant. Dose, 1 to 4.	40	1 75
Codein. ¼ gr........................ Med. prop.—Anodyne, replacing morphia without the usual disagreeable after-effects produced by the latter. Dose, 2 to 4.	1 25	6 00
Corrosive Sublimate. 1-12, 1-20, 1-40, and 1-100 gr................. Med. prop.—Mercurial alterative. Dose, 1 to 2.	40	1 75

GRANULES.	BOTTLE.	
	100	500
Digitalin. (Alkaloid.) 1-60 gr...... Med. prop.—Arterial sedative. Dose, 1 to 2.	75	3 50
Elaterium. (Clutterbuck's) 1-10 gr. Med. prop.—Diuretic, Hydragogue cathartic. Dose, 1 to 2.	95	4 50
Ext. Belladonna. (English) ¼ gr Med. prop.—Anodyne. Dose, 1 to 3.	40	1 75
Extract Cannabis Indica. ¼ gr. Med. prop.—Anodyne. Dose, 1 to 4.	60	2 75
Extract Hyoscyam. (Eng.) ½ gr. Med, prop.—Nerve stimulant. Dose, 1 to 3.	40	1 75
·Extract Ignatia Amara. ¼ gr.. Med. prop.—Nerve sedative. Dose, 1 to 2.	50	2 25
Extract Nuc. Vom. ¼ and ½ gr.. Med, prop.—Nerve stimulant. Dose, 1 to 3.	40	1 75
Gelsemin. 1-16 gr................. Med. prop.—Arterial sedative. Dose, 1 to 4.	40	1 75
Gelsemin. ⅛ gr................. Med. prop.—Arterial sedative. Dose, 1 to 2.	50	2 25
Gelsemin. ¼ gr................. Med. prop.—Arterial sedative. Dose, Dose, 1 to 2.	75	3 50
Helonin. 1-10 gr................. Med. prop.—Cathartic. Dose, 1 to 4.	50	2 25
Hydrastin. ½ gr................. Med. prop.—Cathartic. Dose, 1 to 2.	95	4 50
Hyoscyamiæ. 1-100 gr............ (Crystals, Pure Alkaloid.) Med. prop.—Anodyne, Soporific. Dose, 1.	3 00	14 75
Leptandrin. ¼ gr................ Med. prop.—Cathartic. Dose, 1 to 4.	40	1 75
Leptandrin. ½ gr................ Med. prop.—Cathartic. Dose, 1 to 4.	50	2 25
Mercury Bin-Iodide. 1-16 gr.... Med. prop.—Anodyne. Dose, 1 to 4.	40	1 75

52

GRANULES.	BOTTLE. 100	BOTTLE. 500
Mercury Protiodide. ¼ gr....... Med. prop.—Alterative. Dose, 1 to 4. Prepared from the precipitated Iodide and free of all traces of the Bin-Iodid.	40	1 75
Mercury Protiodide. ½ gr....... Med. prop.—Alterative. Dose, 1 to 2.	50	2 25
Mercury Protiodide. ⅛ gr....... Med. drop.—Alterative. Dose, 2 to 4.	40	1 75
Mercury Protiodide. 1-5 gr....... Med. prop.—Alterative. Dose, 1 to 4.	40	1 75
Morphine Acet. ⅛ gr.............. Med. prop.—Anodyne. Dose, 1 to 2.	50	2 25
Morphine Sulph. 1-20 gr......... Med. prop.—Anodyne. Dose, 1 to 2.	40	1 75
Morphine Sulph. 1-10 gr......... Med. prop.—Anodyne. Dose, 1 to 2.	40	1 75
Morphine Sulph. ⅛ gr........... Med. prop.—Anodyne. Dose, 1 to 2.	50	2 25
Morphine Sulph. ⅙ gr........... Med. prop.—Anodyne. Dose, 1 to 2.	60	2 75
Morphine Sulph. ¼ gr........... Med. prop.—Anodyne. Dose, 1 to 2.	75	3 50
Morphine Sulph. ½ gr........... Dose, 1.	1 25	6 00
Morphine Valer. ⅛ gr........... Med. prop.—Anodyne. Dose, 1 to 2,	90	4 25
Podophyllin. 1-10 gr............... Med. prop.—Cathartic. Dose, 1 to 4	40	1 75
Podophyllin. ¼ gr.................. Med. prop.—Cathartic. Dose, 1 to 4	40	1 75
Podophyllin. ½ gr .... Med. prop.—Cathartic. Dose, 1 to 2.	50	2 25
Podophyllin Comp............... Med. prop.—Cathartic and Tonic. Dose, 1 to 2. Podophyllin, ¼ gr. Ext. Hyoscyami, ⅛ gr. Ext. Nuc. Vom., 1-16 gr.	75	3 50

GRANULES.	BOTTLE.	
	100	500
Silver Iodid. ¼ gr................ Med. prop.—Alterative to Mucous Membrane. Dose, 1 to 4.	75	3 50
Silver Nitrate. ¼ gr............. Med. prop.—Alterative to Mucous Membrane. Dose, 1 to 4.	75	3 50
Strychnin. 1-16, 1-20, 1-30, 1-32, 1-40, and 1-60 gr................ Med. prop.—Nerve stimulant, Tonic. Dose, 1 to 3.	40	1 75
Strychninæ Sulph. 1-32......... Med. prop.—Tonic. Dose, 1 to 2.	40	1 75
Veratrinæ Sulph. 1-12 gr....... Med. prop.—Powerful topical excitant. Dose, 1.	50	2 25
Zinc Phosphide. ⅛ and ¼ gr.... Med. prop.—Tonic. Dose, 1 to 3.	75	3 50

Antiseptic Pastilles.

(Dr. Carl Seiler's Formula.)

Useful in Nasal Catarrh and Inflamed Mucous Membranes. Used as an Antiseptic Spray, Nasal and Mouth Wash.

DIRECTIONS.—Dissolve one tablet in two fluid ounces of warm water to be sniffed up the nose or used as a spray or with a soft tooth brush, as a mouth wash for tender gums. It is a valuable antiseptic and healing agent, imparting a pleasant and cleansing after effect when used as directed. Per bottle, 50 cts.

PREPARED BY

WILLIAM R. WARNER & CO.

PHILADELPHIA AND NEW YORK.

ADDENDA

SOLUBLE COATED PILLS

PILLS.	BOTTLE. 100	500
Euonymin. 2 grs.................... Med. prop.—Tonic, Laxative, Diuretic. Dose, 1 to 2.	2 00	9 75
Cocaine Hydrochlor. 1-10 gr... Med. prop.—Stimulant, Tonic, Aphrodisiac.	1 50	7 25
Ergotin. 2 grs................. Med. prop.—Parturient. Dose, 1 to 2.	1 25	6 00
Ext. Eucalyptus. 2 grs........... Med. prop.—Diaphoretic, Febrifuge.	1 00	4 75
Morphinæ Murias. ½ gr........ Med. prop.—Anodyne, Soporific. Dose, 1.	1 50	7 25
Morphinæ Murias. ¼ gr........ Med. prop.—Anodyne, Soporific.	90	4 25
Morphinæ Murias. ⅛ gr........ Med. prop.—Anodyne, Soporific. Dose, 1 to 2.	60	2 75
Nickel Bromid. 2½ grs.......... Med. prop.—Recommended in Epilepsy. Dose, 1 to 2.	1 50	7 25
Sumbul Comp....... .... Ext. Sumbul, 1 gr. Asafœtida, 2 grs. Ferri Sulph. Exsic., 1 gr. Acid Arsen., 1-30 gr.	1 50	7 25

PILLS.	BOTTLE.	
	100	500

Quinina et Ferri et Zinci Valer........................... 1 50 7 25
Med. prop.—Recommended for the re-
lief of Melancholia, Incipient Insanity,
and as being particularly adapted for the
cure of the worry of nervous females.
Dose, 1.
R Quininæ Valer.
Ferri Valer.
Zinci Valer. āā 1 gr.

Chalybeate Comp. (Warner & Co.)   80
Med. prop.—Employed in the treatment
of Anemia, Chlorosis, Phthisis, etc. Dose,
1 to 3.
Chalybeate Mass.   2 grs. }
Ext. Nuc. Vom.   ⅛ gr. }

Manganese Bin-Oxide. 2 grs... 1 25 6 00
Med. prop.—Emmenagogue. Dose, 1.

Aloin. ¼ gr..........................   75 3 50
Med. prop.—Laxative. Dose, 1 to 3.

Antiseptic. (Warner & Co.)   80
Med. prop.—Pil. Antiseptic is prescribed with great advantage in cases of
Dyspepsia attended with Acid Stomach
and Enfeebled Digestion following an
over indulgence in eating or drinking. It
is also useful in Rheumatism. Dose,
1 to 3.
Sulphite Soda,   1 gr. }
Salicylic Acid,   1 gr. }
Ext. Nuc. Vom.   ¼ gr. }

Antiseptic Comp. (Warner & Co.)   80
Med. prop.—Used with great advan-
tage in cases of Dyspepsia, Indigestion
and Malassimilation of food. Dose, 1
to 3.
Sulphite Soda,   1 gr. }
Salicylic Acid,   1 gr. }
Ext. Nuc. Vom.   ⅛ gr. }
Powd. Capsicum,   1-10 gr. }
Concent. Pepsin,   1 gr. }

Lapactic.......................   60 2 75
Aloin,   ¼ gr. }
Strychnin,   1-60 gr. }
Ext. Belladon.   ⅓ gr. }
Ipecac,   1-16 gr. }

56

PILLS	BOTTLE	
	100	500
Salutis, (Dr. T. G. Thomas)........	80	3 75
Aloin, ½ gr. Ext. Hyoscy, ½ gr. Ext. Nuc. Vom ⅛ gr. Ext. Belladon. 1-16 gr.		
Tonic-Laxative, (Dr. Skene).....	80	3 75
Dose, 1. Quininæ Sulph. 1 gr. Ext. Colocynth. Co. ½ gr Ext. Belladon. 1-10 gr.		
Ferruginous, (Blaud) Improved. Dose, 1.	60	2 75
Ferruginous, (Blaud) 3 grs....... Dose, 1 to 4.	40	1 75
Ferruginous, (Blaud) 5 grs. ...... Dose, 1 to 2.	50	2 25
Ferri Carb. (Vallett) 5 grs........ Dose, 1 to 2.	50	2 25
Ext. Damiana, 2 grs.............. Dose, 1.	1 00	4 75
Ext Damiana, 3 grs.............. Dose, 1.	1 00	4 75

WM. R. WARNER & CO.

MANUFACTURING CHEMISTS,

PHILADELPHIA and NEW YORK.

A NEW
Important Class of Remedies.
PARVULES.

☞ YOU ARE CAUTIONED AGAINST IMITATIONS
AND SUBSTITUTIONS OFFERED UNDER
OTHER NAMES.

This is a new class of medicines (minute pills), designed for the administration of remedies in small doses for frequent repetition in cases of children and adults. It is claimed by some practitioners that small doses, given at short intervals, exert a more salutary effect. *The elegence and efficiency of Parvules, and the avoidance of cumulative effect, depend on our mode of preparation.*

THE DOSE
of any of the parvules will vary from one to four, according to age or the frequency of their admistration. For instance, one Parvule every hour, or two every two hours, or three every three hours, and so on for adults. For children, one three times a day is the minimum dose.

PRICE, 25 CENTS PER BOTTLE OF 100 EACH. DISCOUNT FOR QUANTITIES.

POCKET CASES, WITH 20 VARIETIES, FOR THE USE OF PRACTITIONERS, $5.00 NET.

POCKET CASES, WITH 10 VARIETIES, FOR THE USE OF PRACTITIONERS, $2.50.

HAND OR BUGGY CASES, 40 BOTTLES, ALL THE VARIETIES, $10.00 NET.

SUPPLIED BY ALL DRUGGISTS, OR SENT BY MAIL ON RECEIPT OF PRICE.

Acidi Arseniosi.......................1-100 gr.
Medical properties.—Alterative, Anti-periodic.

Acidi Salicylic......................1-10 gr.
Med. prop.—Anti-rheumatic.

Acidi Tannic....................1-20 gr.
  Med. prop.—Astringent.

Aconiti Rad.................1-20 gr.
  Med. prop.—Narcotic, Sudorific.

Aloin............................1-10 gr.
  Med. prop.—A most desirable cathartic.

Dose.—4 to 6 at once. This number of Parvules, taken at any time, will be found to exert an easy, prompt, and ample cathartic effect, unattended with nausea, and in all respects furnishing the most desirable aperient and cathartic preparation in use. For habitual constipation they replace, when taken in single Parvules, the various medicated waters without the quantity which they require as a dose, which fills the stomach and deranges the digestive organs.

Ammonii Chlorid.................1-10 gr.
  Med. prop.—Diuretic, Stimulant.

Antimonii et Potass. Tart.........1-100 gr.
  Med. prop.—Expectorant, Alterative.

Arnicæ Flor............................-5 gr.
  Med. prop.—Narcotic, Stimulant, Diaphoretic.

Arsenici Iodid...................1-100 gr.
  Med. prop.—Alterative.

Belladonnæ Fol...................1-20 gr.
  Med. prop—Narcotic, Diaphoretic, Diuretic.

Calomel.............................1-20 gr.
  Med. prop.—Alterative, Purgative.

Dose.—1 to 2 every hour. Two Parvules of calomel, taken every hour until five or six doses are administered (which will comprise but half a grain), produce an activity of the liver which will be followed by bilious dejections and beneficial effects that twenty grains of blue mass or ten grains of calomel rarely cause, and sickness of the stomach does not usually follow.

Calomel et Ipecac................ āā 1-10 gr.
  Med. prop.—Alterative, Purgative.

Camphoræ........................1-20 gr.
  Med. prop.—Diaphoretic, Carminative.

Cantharidis......................1-50 gr.
  Med. prop.—Diuretic, Stimulant.

Capsici..................................1-20 gr.
Med. prop.—Stimulant and Carminative.

Cathartic Comp., Officinal............⅔ gr.
Med. prop.—Cathartic.

Cathartic Comp., Improved........⅓ gr.
Med. prop.—Cathartic.

Digitalis Fol............................1-20 gr.
Med. prop.—Sedative, Narcotic, Diuretic.

Dover's Powder........................⅓ gr.
Med. prop.—Anodyne, Soporific.

Ergotin..................................1-10 gr.
Med. prop.—Emmenagogue, Parturient.

Ferri Redact............................1-10 gr.
Med. prop.—Tonic.

Gelsemini Rad..........................1-50 gr.
Med. prop.—Nervous and Arterial Sedative.

Hydrarg. Bi-Chlor......................1-100 gr.
Med. prop.—Mercurial alterative, Germicide.

Recently, Corros. Sub., in small doses, has been administered with most gratifying results in certain forms of Chronic Dyspepsia.

Dose.—One Parvule, repeated according to the age or nature of the disease. Ringer and other eminent therapeutists extol very highly 1-100 gr. in dysentery of children, regarding it as almost *specific.*

Hydrarg. cum Creta....................1-10 gr.
Med. prop.—Alterative.

Hydrarg. Iodid..........................1-20 gr.
Med. prop.—Alterative.

Hydrastin................................1-20 gr.
Med. prop.—Tonic, Astringent.

Iodoform................................1-10 gr.
Med. prop.—Alterative.

Ipecac..................................1-50 gr.
Med. prop.—Emetic, Expectorant.

Morphinæ Sulph........................1-50 gr.
Med. prop.—Narcotic, Sedative.

60

Nucis Vomicæ...................1-50 gr.
Med. prop.—Tonic, Stimulant.

Opii........................1-40 gr.
Med. prop.—Narcotic, Sedative, Anodyne.

Phosphorus...................1-200 gr.
Med. prop.—Nerve stimulant.

Piperin......................1-20 gr.
Med. prop.—Tonic, Anti-periodic, Carminative.

Podophyllin..................1-40 gr.
Med. prop.—Carhartic, Cholagogue.
Two parvules of podophyllin administered three
times a day will re-establish and regulate the peri-
staltic action and relieve habitual constipation, add
tone to the liver, and invigorate the digestive functions.

Potass. Arsenitis.............1-100 gr.
Med. prop.—Alterative.

Potass. Bromid...............1-5 gr.
Med. prop.—Alterative, Resolvent.

Potass. Nitratis.............1-10 gr.
Med. prop.—Diuretic and Refrigerant.

Quininæ Sulphas..............1-10 gr.
Med. prop.—Tonic, Anti-periodic.

Santonin.....................1-10 gr.
Med. prop.—Anthelmintic.

Strychninæ...................1-100 gr.
Med. prop.—Nerve stimulant, Tonic.

PREPARED ONLY BY

WILLIAM R. WARNER & CO.,

1228 MARKET STREET,

PHILADELPHIA.

SOLUBLE
Hypodermic Tablets.

In compliance with repeated requests from the medical profession to manufacture Tablets for subcutaneous medication, we have prepared a series as per following list, which will be found easily and quickly soluble.

PRICE PER BOX, ASSORTED, 12 TUBES, 20 TABLETS IN EACH, $2.35.

	Per tube.	Per doz.
PILOCARPINÆ HYDROCHL., 1-5 gr.	$0.20	$2.00
ERGOTIN, 1-6 gr.	.20	2.00
APOMORPHINE, 1-12 gr.	.20	2.00
ATROPINÆ SULPH., 1-60 gr.	.20	2.00
STRYCHNINÆ SULPH., 1-60 gr.	.20	2.00
MORPHINÆ SULPH., ¼ gr. et ATROPINÆ SULPH., 1-120 gr.	.25	2.50
MORPHINÆ SULPH., ⅓ gr.	.20	2.25
MORPHINÆ SULPH., ¼ gr.	.20	2.25
MORPHINÆ SULPH., 1-6 gr.	.15	.75
HYOSCYAMIN SULPH., 1-100 gr.	.60	5.75
HYDRARG. BI-CHLOR., 1-150 gr.	.20	2.00
ESERINÆ SULPH., 1-60 gr.	.30	3.00

THE VIALS SHOULD BE KEPT TIGHTLY CORKED.

DR. GRANT'S
IMPROVED HYPODERMATIC SYRINGE.

PRICE, $3.50 EACH.

ILLUSTRATED CIRCULAR SENT ON APPLICATION.

Globules sent by mail on receipt of price.

WM. R. WARNER & CO.

MANUFACTURING CHEMISTS,

PHILADELPHIA AND NEW YORK.

62

DOSIMETRIC GRANULES.

WM. R. WARNER & CO.

As their name indicates, these Granules are *measured doses* of the alkaloids, metals and metalloids in such definite and accurate pro-portions as may best meet the requirements of the physician.

The most perfect system of Dosimetry is that comprised in Parvules originally introduced by Warner & Co. several years ago, but Dosimetric Granules are intended to comprise such reme-dies as are new and such as are *proximate prin-ciples* not so frequently repeated in a measured time for all cases.

These Granules have been divided according to the metric system into strengths of half mil-ligramme, one milligramme and centigramme. In each instance, however, their equivalents are stated in *grains* or fractions thereof. Such a plan, we think will easily familiarize the practitioner with the metric system for all practical purposes and will commend itself at once to his recognition.

The selection of the list is based entirely upon the *physiological action* of these active or essen-tial principles, and are therefore administered upon a rational basis, thus making therapeu-tics stand upon firmer ground than that which it has occupied or even still occupies, when compared with the advances which its sister sciences have taken within the last decade. Under the administration of these Granules, therapeutics partake of the character of a *specific* form of treatment.

The cutting short or strangulation of many acute diseases while as yet in their incipient or formative stages has not been sufficiently appreciated. That this is possible, the medical literature of the day affords ample evidence; but to accomplish it, treatment must be both scientific and energetic, i. e., must be based upon the physiological action of drugs and upon the action of reliable medications.

With such an intention, these Granules have been prepared abroad, (in France particularly as suggested by Dr. Burggraeve.)

This method of treatment has met with success, and it will be a matter of no surprise that therapeutists in this country should be prompt in adopting it.

In the hands of the physician and *his hands only*, these Granules are potent remedies, capable of accomplishing results far more quickly and certainly than the uncertain fluid extracts and tinctures, and far more pleasantly.

See following pages for list comprising 66 varieties.

PREPARED BY

WM. R. WARNER & CO.

MANUFACTURING CHEMISTS.

PHILADELPHIA AND NEW YORK.

DOSIMETRIC GRANULES.

The dose is one Granule, repeated to meet the requirements of the case.

Per 100

Aconitin..............1-65 gr. (1 milligram)$ 75
Med. prop.—Nerve sedative.

Acid Arseniosum....1-65 gr. (1 milligram) 40
Med. prop.—Anti-periodic, Alterative.

Acid Phosphoric.....1-65 gr. (1 milligram) 40
Med. prop.—Nerve stimulant.

Acid Tannicum........⅙ gr. (1 centigram) 40
Med. prop.—Astringent.

Acid Salicylicum......⅙ gr. (1 centigram) 40
Med. prop.—Anti-rheumatic.

Antimonii Arsenias..1-65 gr. (1 milligram) 40
Med. prop.—Alterative, Diaphoretic.

Arsenii Iodidum.....1-65 gr. (1 milligram) 40
Med. prop.—Alterative.

Asparagin............1-65 gr. (1 milligram) 40
Med. prop.—Arterial sedative.

Atropinæ Sulphas.1-130 gr. (½ milligram) 75
Med. prop.—Anodyne, Anti-spasmodic.

Brucin..............1-130 gr. (½ milligram) 40
Med. prop.—Tonic.

Bryonin.............1-65 gr. (1 milligram) 50
Med. prop.—Hydragogue, Cathartic.

Caffeinæ Arsenias....1-65 gr. (1 milligram) 50
Med. prop.—Alterative.

Caffeinæ Citras.......1-65 gr. (1 milligram) 50
Med. prop.—Nerve stimulant.

Caffeinæ Valerianas 1-65 gr. (1 milligram) 50
Med. prop.—Stimulant, Anti-spasmodic.

Calabarin. Sulphas.1-130 gr. (½ milligram) 75
Med. prop.—Spinal sedative.

Calomel..............1-65 gr. (1 milligram) 40
Med. prop.—Alterative, Purgative.

Camphoræ Bromated.⅙ gr. (1 centigram) 40
Med. prop.—Sedative.

GRANULES ARE NOT PARVULES.

Per 100

Cicutin..................1-130 gr. (½ milligram) 50
Med. prop.—Nerve sedative.

Cicutin. Hydrobromas. . . 1-65 gr.
(1 milligram) 75
Med. prop.—Nerve sedative.

Codein..................1-65 gr. (1 milligram) 75
Med. prop.—Hypnotic sedative.

Colchicin..................1-130 gr. (½ milligram) 75
Med. prop.—Sedative, Diuretic, Emetic.

Croton Chloral........⅛ gr. (1 centigram) 75
Med. prop.—Hypnotic.

Cubebin..................1-65 gr. (1 milligram) 50
Med. prop.—Diuretic.

Daturin..................1-130 gr. (½ milligram) 75
Med. prop.—Narcotic, Anodyne.

Diastase..................⅛ gr. (1 centigram) 75
Med. prop.—Possesses the power of converting starch into sugar (of the grape.)

Elaterin..................1-65 gr. (1 milligram) 75
Med. prop.—Purgative.

Emetine..................1-65 gr. (1 milligram) 75
Med. prop-Emetic, Diaphoretic, Expectorant.

Ergotin..................⅛ gr. (1 centigram) 40
Med. prop.—Emmenagogue, Parturient.

Ferri Arsenias........1-65 gr. (1 milligram) 40
Med. prop.—Tonic, Alterative.

Ferri Salicylas........ ⅛ gr. (1 centigram) 40
Med. prop.—Tonic.

Ferri Valerianas......⅛ gr. (1 centigram) 40
Med. prop.—Tonic, Anti-spasmodic

Hydrargyri Iodid. Rub..1-65 gr.
(1 milligram) 40
Med. prop.—Alterative.

Hydrargyri Iodid, Vir.⅛ gr. (1 centigram) 40
Med. prop.—Alterative.

Hyoscyamin........1-130 gr. (½ milligram) 3 00
Med. prop.—Hypnotic, Anti-spasmodic.

GRANULES ARE NOT PARVULES.

Per 100

Iodoform............1-65 gr. (1 milligram) 40
Med. prop.—Alterative.

Jalapin...............1-65 gr. (1 milligram) 50
Med. prop.—Hydragogue cathartic.

Koosin.................1-65 gr. (1 milligram) 50
Med. prop.—Anthelmintic.

Lithii Carbonas.......½ gr. (1 centigram) 50
Med. prop.—Diuretic.

Lithii Benzoas........½ gr. (1 centigram) 75
Med. prop.—Diuretic, Expectorant.

Morphinæ Hydrobromas..1-65 gr.
(1 milligram) 75
Med. prop.—Anodyne.

Morphinæ Iodohydras....1-65 gr.
(1 milligram) 75
Med. prop.—Anodyne.

Narcein...............1-65 gr. (1 milligram) 75
Med. prop.—Supposed to influence the inferior part of the spinal marrow, diminishing
the sensation and mobility in the inferior
extremities.

Piperina.............1-65 gr. (1 milligram) 40
Med. prop.—Local and general stimulant.

Picrotoxin...........1-130 gr. (1 milligram) 75
Med. prop.—Narcotic.

Pilocarpin..........1-65 gr. (1 milligram) 75
Med. prop.—Sudorific.

Podophyllin.........½ gr. (1 centigram) 40
Med. prop.—Cholagogue cathartic.

Potasii Arsenias.....1-65 gr. (1 milligram) 40
Med. prop.—Alterative.

Quassin...............1-65 gr. (1 milligram) 50
Med. prop.—Tonic, Febrifuge, Anthelmintic.

Quiniæ Arsenias..1-65 gr. (1 milligram) 50
Med. prop.—Tonic, Alterative.

Quiniæ Hydrobromas....½ gr.
(1 centigram) 75
Med. prop.—Tonic, Anti-spasmodic.

GRANULES ARE NOT PARVULES.

	Per 100
Quininæ Hydroferrocyanas 1-65 gr. (1 milligram) Med. prop.—Tonic.	75
Quininæ Salicylas....⅙ gr. (1 centigram) Med. prop.—Tonic, Stimulant.	75
Quininæ Sulphas.....⅙ gr. (1 centigram) Med. prop.—Tonic, Anti-periodic.	75
Quininæ Valerianas..⅙ gr. (1 centigram) Med. prop.—Tonic, Anti-spasmodic.	75
Santonin................⅙ gr. (1 centigram) Med. prop.—Anthelmintic.	40
Scillitin................1-65 gr. (1 milligram) Med. prop.—Cardiac sedative, Diuretic.	50
Sodii Benzoas.........⅙ gr. (1 centigram) Med. prop.—Diaphoretic, Expectorant.	40
Sodii Salicylas........⅙ gr. (1 centigram) Med. prop.—Hepatic stimulant.	40
Strychninæ Arsenias....1-130 gr. (½ milligram) Med. prop.—Tonic, Alterative.	50
Strychninæ Hypophos...1-130 gr. (½ milligram) Med. prop.—Tonic.	50
Strychninæ Sulphas....1-130 gr. (½ milligram) Med. prop.—Tonic.	40
Sulphur Iodidum.....⅙ gr. (1 centigram) Med. prop.—Alterative.	40
Veratrina.............1-130 gr. (½ milligram) Med. prop.—Topical excitant.	40
Zinci Cyanidum.....1-65 gr. (1 milligram) Med. prop.—Anti-spasmodic.	40
Zinci Phosphidum .1-65 gr. (1 milligram) Med. prop.—Tonic, Stimulant.	40
Zinci Valerianas.....⅙ gr. (1 centigram) Med. prop.—Anti-spasmodic.	40

GRANULES ARE NOT PARVULES.

Granular Effervescent Salts.

(EQUAL TO THE BEST ENGLISH.)

PREPARED BY

WILLIAM R. WARNER & CO.

GRANULAR EFFERVESCENT

ANTALGIC SALINE.

Each dessertspoonful contains
ANTIPYRINE, - - - - 4 grains.
SALICYLATE SODA, - - - 4 grains.-

A specific in Neuralgic Headaches.

GRANULAR EFFERVESCENT

ANTIPYRINE.

Each dose or heaping teaspoonful contains
5 grains Antipyrine.

The effervescent form is the most desirable for the
administration of this remedy because of the attending
Carbonic Acid; which according to Prof. Germain,
avoids the tendency to produce nausea and the rash,
which frequently follows the administration of Antipyrine.

GRANULAR EFFERVESCENT,

ACID. SALICYLIC.

Each teaspoonful contains five grains of Salicylic
Acid. Valuable in Rheumatism and analogous disorders.
The temperature should be carefully watched in
special cases.

GRANULAR EFFERVESCENT.

BROMO SODA.

COMPOSITION:

CAFFEINE,	1 gr.
BROMIDE SODIUM,	30 gr.

In each teaspoonful.

Useful in Nervous Headache, Sleeplessness, Excessive Study Over Brainwork, Nervous Debility, Mania, etc., etc.

Dose.—A heaping teaspoonful in half a glass of water, to be repeated once after an interval of thirty minutes, if necessary.

It is claimed by some prominent specialists in nervous diseases, that the Sodium Salt is more acceptable to the stomach than the Bromide Potassium. An almost certain relief is given by the administration of this Effervescing Salt. It is also used with advantage in *Indigestion*, *Depression* following alcoholic and other excesses, as well as *Nervous Headache*. It affords speedy relief for *Mental and Physical Exhaustion*.

GRANULAR EFFERVESCENT

CAFFEINE

AND

BROMIDE OF POTASSIUM.

Specially prepared by William R. Warner & Co.

Dose.—A large teaspoonful, in water containing

HYDROBROMATE OF CAFFEINE,	1 gr.
BROMIDE OF POTASSIUM,	20 grs.

PROPERTIES:—Useful in *Sleeplessness, Over Exertion of the Brain, Over Study, Nervous Debility, etc.* and in all cases for which the above remedies are given singly to advantage. An almost certain relief is given by the administration of this Effervescing Salt. It affords a pleasant and delightful draught, by mixing a large teaspoonful with a glass of water and drinking while effervescing. It is also used with advantage in *Indigestion*, *Depression* following alcoholic and other excesses, as well as *Nervous Headache*. It affords speedy relief for *Mental and Physical Exhaustion*. Physicians recognize its great advantage. The dose named may be repeated, if necessary, three times at intervals of thirty minutes.

GRANULAR EFFERVESCENT

CITRATE OF BISMUTH.

Each dose or heaping teaspoonful of the salt contains two grains of soluble Citrate of Bismuth, and should be placed in part of a glass of water and drunk while effervescing.

GRANULAR EFFERVESCENT

CITRATE OF CAFFEINE.

A dessertspoonful containing one grain of Citrate of Caffeine should be given for sick headache every hour or two before and during the paroxysms.

This preparation should be used more particularly in those cases of sick headache, the cause of which is torpidity of the liver and other excretory organs.

GRANULAR EFFERVESCENT

CITRATE LITHIA AND POTASH.

Each dose or heaping teaspoonful contains
LITHII CITRAS.   5 grains.
POTASSII CITRAS.   10 grains.

To be taken in a glass of water at a temperature not under 60°

GRANULAR EFFERVESCENT

CRAB ORCHARD SALT.

Prepared from the exact of analysis of the purified natural salt as obtained from the Crab Orchard Spring, Ky.

A heaping teaspoonful in half of a glass of water forms a grateful, effervescent draught, producing the effect of the natural water.

GRANULAR EFFERVESCENT

HYDROBROMATE CAFFEINE.

A dessertspoonful containing one grain of Hydrobromate of Caffeine, should be given for sick headache every hour or two before and during the paroxysms.

This most valuable therapeutic agent acts with marvelous promptitude in those cases of headache the cause of which is gastric irritability from over indulgence in food or drink, and also when due to nervous exhaustion caused by want of sleep and overwork.

71

GRANULAR EFFERVESCENT

NITRATE OF CERIUM.

Each heaping teaspoonful contains two grains of Nitrate of Cerium.

Useful in reflex vomiting and gastric distress not due to organic disease of the stomach.

The dose may be increased to four and even eight grains of Nitrate of the Cerium as occasion demands.

GRANULAR EFFERVESCENT

CONGRESS SALT.

A Saline Aperient.

A heaping teaspoonful in a glass half full of water, forms an agreeable and refreshing draught identical with the water of the popular "Congress Spring" at Saratoga.

GRANULAR EFFERVESCENT

OXALATE OF CERIUM.

Each heaping teaspoonful contains two grains of Oxalate of Cerium.

Valuable in *reflex* vomiting not due to gastric irritability. In many cases of so-called nervous Dyspepsia it will control the paroxysms.

GRANULAR EFFERVESCENT

GINGERADE.

A tablespoonful dissolved in a glass of water forms an exceedingly grateful summer drink, similar to Ginger Ale.

GRANULAR EFFERVESCENT

KISSINGEN SALT.

A heaping teaspoonful dissolved in half of a glass of water, forms an agreeable and refreshing draught, identical with the natural water.

GRANULAR EFFERVESCENT

BENZOATE OF LITHIA.

Each teaspoonful contains four grains of chemically pure salt.

This preparation has been strongly recommended as a remedy for gout. It may be used with good effect in all cases of Lithaemia, Gout and Rheumatic Gout.

72

GRANULAR EFFERVESCENT

BROMIDE OF LITHIA.

Each teaspoonful contains five grains of the chemically pure salt.

This preparation has been strongly recommended as a remedy for epilepsy and as a hypnotic of great value.

GRANULAR EFFERVESCENT

SALICYLATE OF LITHIA.

DOSE.—A teaspoonful, containing ten grains of the salt. A convenient and pleasant remedy in Gout and Rheumatism.

This preparation is intended for physicians' use, and will be found to possess advantages over Salicylic Acid, being less irritating to the stomach, and combining the efficacy of Lithia and Salicylic Acid.

GRANULAR EFFERVESCENT

CITRATE LITHIA.

Each heaping teaspoonful contains four grains of the chemically pure salt. Valuable in Rheumatic, Gouty, and analogous disorders being acceptable to delicate stomachs, where the Carbonate is not well borne.

GRANULAR EFFERVESCENT

CARBONATE LITHIA.

Each heaping teaspoonful contains four grains of the chemically pure salt. A remarkable and often magical resolvent of Gouty Rheumatic deposits.

Dr. A. Garrod, the well-known English authority on Gout, and who was the first physician to introduce the Lithia Salts in the treatment of the gouty diathesis, states that their action is materially increased by being administered in a *freely diluted form*.

The effervescing salts of Lithia furnish an easy and elegant way of applying Dr. Garrod's methods.

GRANULAR EFFERVESCENT
CHALYBEATE SALINE.
(FERRIC SALINE EFFERVESCENS, DR. MEANS.)

Each heaping teaspoonful contains one grain of Citro-tartrate of Iron and twenty grains of Soda.

Dose.—A heaping teaspoonful of the Salt, containing one grain of Citro-tartrate of Iron and twenty grains of Soda, to be taken in a glass two-thirds full of water and drunk while effervescing. If a more decided effect is desired, warm instead of cold water may be used. In all cases this draught should be taken but once or twice a day, and then on an empty stomach, preferably before breakfast. No restrictions as to diet. One or two Pil. Digestiva (W.& Co.) may be taken at noon, before eating as a dinner pill.

GRANULAR EFFERVESCENT
CITRATE OF MAGNESIA.
Pleasant and Efficient.

DIRECTIONS.

For a purgative effect take two or more tablespoonfuls added to a small glass of water, and drink while effervescing. As a laxative, one or two tablespoonfuls, taken in the same manner. One or two teaspoonfuls in sweetened water, produces a delightful cooling drink in summer.

This preparation of Citrate of Magnesia has the advantage over the liquid form, and a fresh and effective preparation is possible in every household.

GRANULAR EFFERVESCENT
SULPHATE OF MAGNESIA.

DIRECTIONS.

For a purgative effect, take two or more tablespoonfuls added to a small glass of water, and drink while effervescing. As a laxative, one or more tablespoonfuls taken in the same manner.

This preparation is more agreeable and quite as efficient as the old fashioned dose of salts, and is thus a great boon to the nursery.

GRANULAR EFFERVESCENT
PEPSIN AND BISMUTH.

Each heaping teaspoonful contains the dose of Saccharated Pepsin and soluble Citrate of Bismuth.

Useful in cases of Indigestion due to catarrhal conditions of the gastric mucous membrane.

74

GRANULAR EFFERVESCENT

PEPSIN, BISMUTH AND IRON.

Each heaping teaspoonful contains the dose of Saccharated Pepsin, soluble Citrate of Bismuth and one grain of Citrate of Iron.

Useful in cases of Dyspepsia, Indigestion and gastric irritability.

GRANULAR EFFERVESCENT

PEPSIN, BISMUTH AND STRYCHNINE.

Each heaping teaspoonful contains the dose of Saccharated Pepsin, soluble Citrate of Bismuth, and the one-sixtieth grain of Citrate of Strychnine.

A valuable and agreeable tonic, indicated in cases of atonic dyspepsia.

GRANULAR EFFERVESCENT

LITHIATED POTASH.

Each heaping teaspoonful contains five grains of Carb. Lithia and ten grains of Bi-Carb. Potash.

GRANULAR EFFERVESCENT

BI-CARBONATE OF POTASSIUM.

A dessertpoonful in a glass of water makes an excellent neutral mixture and effervescent draught The diaphoretic and refrigerant effects are readily produced.

GRANULAR EFFERVESCENT

IODIDE OF POTASSIUM.

Each heaping teaspoonful contains two grains of Iodide of Potassium.

This preparation of Iodide of Potassium has the advantage of being easily borne by the stomach, and as the peculiar taste is disguised, can be taken for a long time without arousing disgust in the patient which is often found when other preparations of Iodide of Potassium have to be taken for a long period.

GRANULAR EFFERVESCENT

BROMIDE OF POTASSIUM.

Each heaping teaspoonful contains ten grains of Bromide of Potassium.

GRANULAR EFFERVESCENT

CITRATE OF POTASSIUM.

A dessertspoonful In a glass of water makes an excellent neutral mxiture and effervescent draught. The diaphoretic and refrigerant effects are readily produced.

GRANULAR EFFERVESCENT

NITRATE OF POTASSIUM.

A dessertspoonful in half of a glass of water makes an excellent mixture and effervescent draught. The diaphoretic and refrigerant effects are readily produced.

GRANULAR EFFERVESCENT

PULLNA SALTS.

A large teaspoonful dissolved in a glass half full of water, forms an agreeable and refreshing draught, identical with the water of the Pullna Springs in Germany.

GRANULAR EFFERVESCENT

ROCHELLE SALTS.

A pleasant and gentle laxative; its use is recommended when a mild purgative effect is desired.
The dose is a heaping dessertspoonful in a glass of water to be drank while effervescing. The dose to be increased to a tablespoonful if required.

GRANULAR EFFERVESCENT

SEIDLITZ MIXTURE.

(Seidlitz Powder, U. S. P.)

An excellent aperient and refrigerant, very acceptable to the stomach.
DOSE.—One tablespoonful in a glass of water.
Like the Citrate of Magnesia this preparation is a more convenient and agreeable form of an aperient medicine than is usually administered.

GRANULAR EFFERVESCENT

BROMIDE OF SODIUM.

Each large teaspoonful contains ten greins of Bromide of Sodium.

GRANULAR EFFERVESCENT

SELTZER SALT.

A tablespoonful to half of a glass of water, forms a grateful and refreshing saline draught, identical with the natural water.

GRANULAR EFFERVESCENT

BI-CARBONATE OF SODIUM.

An excellent antacid, also antilithic and resolvent.
Dose.—A heaping teaspoonful in part of a glass of water.

GRANULAR EFFERVESCENT

PHOSPHATE OF SODIUM.

Each heaping teaspoonful contains thirty grains of Phosphate of Sodium.

GRANULAR EFFERVESCENT

SALICYLATE OF SODIUM.

Anti-Rheumatic.

Each heaping teaspoonful contains ten grains of Salicylate of Sodium.
Salicylate of Sodium is now generally preferred to other form of Salicylic Acid, owing to its greater solubility, etc.

GRANULAR EFFERVESCENT

SALICYLATE OF SODA
With BROMIDE OF POTASH.

Anti-rheumatic, Sedative.

Each heaping teaspoonful contains ten grains of Salicylate of Soda and ten grains of Bromide of Potash.
The dose is usually one large teaspoonful in half of a glass of water, three times a day before eating.
This is the minimum dose for adults, and may be increased with advantage in many cases of Rheumatism and Rheumatic Gout.
This preparation is particularly valuable in cases of Lythiasis, in which the more prominent symptoms are inflammation of the mucous membranes of the respiratory and digestive tracts and ill-defined muscular soreness.

GRANULAR EFFERVESCENT

TRIPLE BROMIDES.

Useful in Headache, Nervousness, Sleeplessness, Migraine, Diurnal Epilepsy, etc., etc.,

DOSE.—A teaspoonful containing Sodium Bromide, 15 grains; Potassium Bromide, 10 grains; 5 grains Ammonium Bromide, three times daily.

Administer one teaspoonful in half of a glass of water. Drink while effervescing. In Diurnal Epilepsy take a dessertspoonful three times daily until sense of taste is partly destroyed. After this reduce the frequency of the dose, but keep the fauces in a benumbed condition.

GRANULAR EFFERVESCENT

VICHY SALT.

(Grand Grille.)

A heaping teaspoonful dissolved in a half of a glass of water, forms a grateful and refreshing draught, identical with the natural water.

GRANULAR EFFERVESCENT

VICHY SALT, LITHIATED.

A large teaspoonful dissolved in half of a glass of water, forms a grateful and refreshing draught, identical with the natural water, combined with the antilithic properties of the Lithia.

PREPARED BY

WILLIAM R. WARNER & CO.,

MANUFACTURING CHEMISTS,

PHILADELPHIA AND NEW YORK.

PARVULE CASES FOR PHYSICIANS' USE.

DELIVERED BY POST ON RECEIPT OF PRICE.

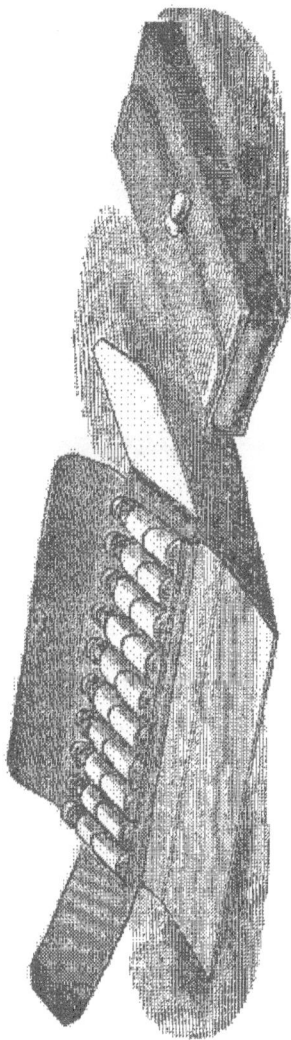

Pocket Parvule Case, filled with ten varieties, any selection, $2.50

This case is of a convenient size for carrying in the coat pocket, and presents a handsome appearance. The dimensions, when closed, are:—Length, 8 in. Width, 3 in. Thickness, 1¼ in.

Manufactured only by WM. R. WARNER & CO.

PARVULE CASES FOR PHYSICIANS' USE.

DELIVERED BY POST ON RECEIPT OF PRICE.

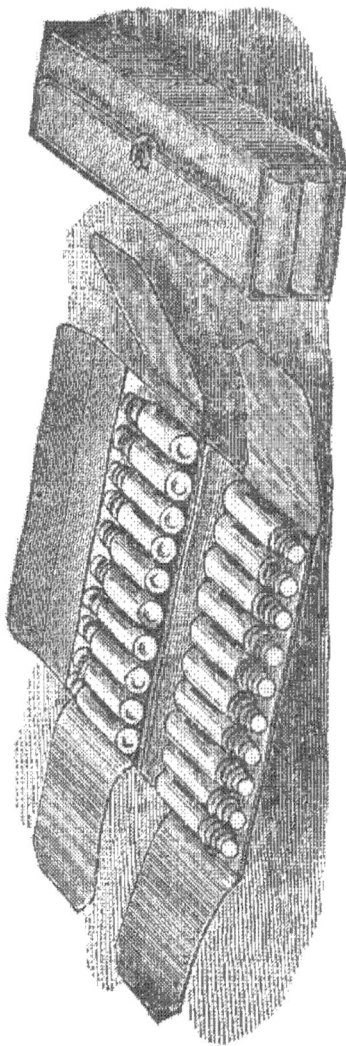

Pocket Parvule Case containing twenty filled bottles, any selection, $5.00

A presentable a. d compact case suitable for pocket or hand. The dimensions, when closed, are :—Length, 8 in. Width, 3 in. Thickness, 2 in.

Manufactured only by **WILLIAM R. WARNER & CO.**

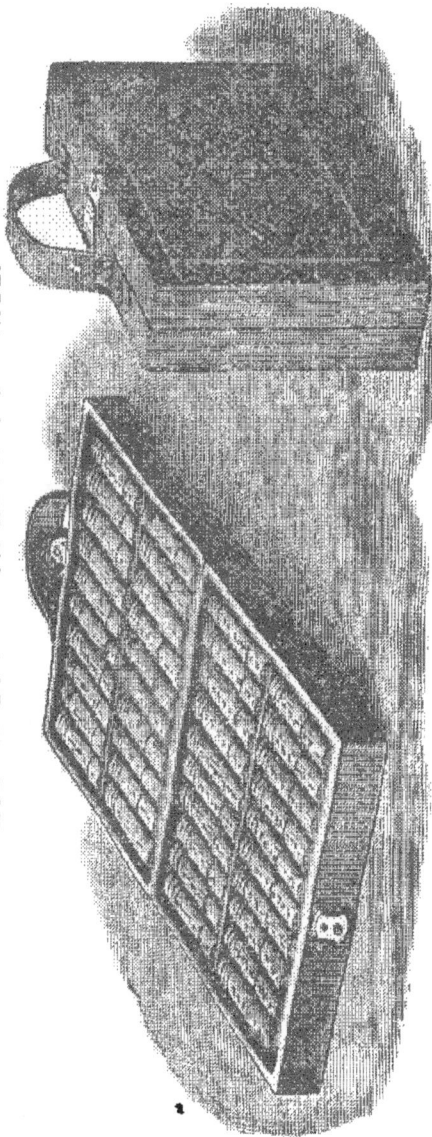

PARVULE CASES FOR PHYSICIANS' USE.

DELIVERED BY POST ON RECEIPT OF PRICE.

Warner & Co.'s Buggy or Hand Parvule Case, containing 40 filled bottles, all the varieties, $10.00.

Manufactured only by WM. R. WARNER & CO.

Office or Stock Case, glass top, containing forty bottles, full variety, $10.00

DELIVERED FREE ON RECEIPT OF PRICE.

This case with glass top is intended as an office case for Physicians and as a stock or show case for Druggists. It is neat and shows the Parvules to good advantage.

Manufactured only by WM. R. WARNER & CO.

"PARVULES."

GENTLEMEN:

Although a practitioner of over forty years, I think I may feel privileged to express my great pleasure and appreciation of the new class of remedies prepared by you called "Parvules." I regard them the greatest improvement in modern medicine, and I could scarcely practice my profession without them, as they are so handy, so convenient, and easily taken by children and adults. Their most important quality is their unvarying and *reliable* strength and efficacy. I can obtain with a grain or less of Calomel, with a grain or less of Aloin, and with a grain or less of Podophyllin, divided respectively into the tenth, twentieth, or fortieth of a grain, in "Parvules," all that I could desire in most cases, and in a more satisfactory manner than in the usual form. I have used successfully a "Parvule" of one-fiftieth of a grain of Sulph. Morphia repeatedly for two or three hours, and have relieved pain without the least nausea or vomiting in patients that could not bear opiates in any other form. I do not know what to attribute this to, except the peculiar mode of preparing the "Parvules," as they are so readily dissolved and absorbed after being taken, and in endorsing them I must disclaim any favoritism or sympathy with Homœopathy. A "Parvule" given every hour, it will be seen, is not Homœopathy in theory or practice. I usually give two "Parvules" of Calomel every hour until six or seven doses are taken, and the result is the same as with ten grains of the same without the embarrassing effect. I give four or five "Parvules" of Aloin, the effect is the same as four or five cathartic pills; also with the Podophyllin "Parvules;" they will relieve habitual constipation, derangement of the liver and digestive organs, if given, one or two, three times a day.

I have no doubt that every practitioner who will use these "Parvules" will find the same results which convinced me of their importance and convenience. I have no other medicine chest in my daily rounds than my pocket case of "Parvules."

Yours very truly,

GEORGE F. REX, M. D.

REAVILLE, N. J.

83

DOSE TABLE

Showing the ordinary mode of adjusting the dose to suit the age of the patient.

The average adult dose being represented by one, the several doses at different ages may be put down as follows.

Age	1 to	3 months, ...	Dose,	1-16
"	4 to 12	"	"	1-10
"	1 to	3 years, ...	"	$\frac{1}{8}$
"	4 to 5	"	. .	"	$\frac{1}{4}$
"	6 to 8	"		"	$\frac{1}{3}$
"	8 to 12	"		"	$\frac{1}{2}$
"	13 to 16	"	. . . .	"	$\frac{2}{3}$
"	17 to 20	"		"	$\frac{3}{4}$
"	21 to 45	"	. . . . .	"	1
"	50		. . . .	"	$\frac{3}{4}$
"	60 to 70	"		"	$\frac{2}{3}$
"	80 to 90	"	. . . .	"	$\frac{1}{2}$
"	100	"		"	$\frac{1}{2}$

The simple rule generally applicable is as follows:

Under 12 years of age diminish the dose of the medicine in the proportion of the age to the age increased by 12 thus:—at 6 years $\frac{6}{6-[-12}=\frac{1}{3}$.

86

EFFERVESCENT

(WARNER & CO.)

SULPHONAL

Each teaspoonful contains 3 grains of Sulphonal.

Sulphonal is the most recent of the hypnotic, and one that has fulfilled the claims made for it. Many eminent medical men, in Europe and in this country, have pronounced in its favor. It has become recognized as an hypnotic of remarkable efficacy, and promises in a great measure to supercede, in the near future, almost all the other sleep producers, such as Chloral, Paraldehyde, etc.

A heaping teaspoonful contains 3 grains of Sulphonal. If repeated every hour until three or four doses are taken, it will produce sleep, from which there will be no unpleasant after effects.

GRANULAR EFFERVESCENT

BROMO
(WARNER & CO.)
LITHIA

Each dessertspoonful contains

Salicylate Lithia,	-	10 grs.
Bromide Soda,	- -	10 grs.

Bromo Lithia is an extremely potent remedy, in the treatment of Rheumatism, Rheumatic Gout, and Gouty Diathesis, originated by W. R. Warner & Co. It consists of Salicylate Lithium, 10 grains, and Bromide Sodium, 10 grains in each dessertspoonful.

It will be found to possess advantages over Salicylic Acid, combining, as it does, the efficacy of Lithium in combination with Salicylic Acid as well as the sedative properties of Bromide of Soda.

Dr. A. Garod, the well known English authority on Gout, who was the first physician to use the Salicylate of Lithia in the treatment of the Gouty Diathesis, believes that its action is materially increased by being administered in a freely diluted form.

Bromo Lithia (Warner & Co.) being an offervescing salt, furnishes an *elegant* and *convenient* form for applying Dr. Garod's methods, and we have pleasure in offering it to thePr ofession. We have attained skill in the manufacture of these elegant effervescent salts, and physicians will receive the benefits of our efforts if they will specify "W. R. W. & Co.'s."

POISONS AND ANTIDOTES.

COMPILED FROM VARIOUS SOURCES.

☞ In all cases use the stomach pump at once if possible.

INORGANIC POISONS.

ACIDS.

Acetic.
Citric.
Muriatic.
Sulphuric.

Nitric.
Oxalic.

Prussic.
Laurel Water.
Nitrobenzole.
Oil Bitter Almond.

Carbonates of sodium, potassium, calcium and magnesium are all antidotes. In the case of sulphuric acid, water should not be drunk, as the union of the two produces great heat. Subsequent inflammation may be treated in the ordinary manner.

Carbonates of calcium and magnesium alone should be employed; see above.

Ammonia is an antidote but it should not be employed in a very concentrated form. Liquid chlorine has also been found efficacious. The cold *douche* to the head has been recommended.

ANTIMONY.

Butter Antim.
Oxide Antim.
Tarter Emetic.

Vomiting should be produced by tickling the fauces and giving large draughts of warm water. Astringent infusions as galls, oak bark, peruvian bark, act as antidotes, and should be given at once. Powdered yellow bark may be given until the infusion is prepared.

ARSENIC.

White Arsenic

Arsenic Acid.

Yellow Arsen.

Emer'd Green.

Hydrated peroxide iron, diffused through water, or the precipitated carbonate in very fine powder, should be given every five or ten minutes until relief is obtained. This is particularly efficacious where white arsenic has been swallowed.

Dialysed iron solution has come much into vogue at the present time and been highly recommended as an antidote, but the recent experiments of E. Hirschsohn, Russia, prove that when used alone, it has no value whatever in this respect, and, when used in connection with ammonia, or magnesia, the resulting insoluble compound formed with the arsenic, is decomposed much more readily in the presence of acids than when the hydrated peroxide of iron is employed.

COPPER.

and Salts.
Verdigris.
Pickles.

Albumen in form most readily obtained, as milk, white of eggs, &c. Vinegar should *not* be given. The inflammatory and nervous symptoms to be treated on general principles.

92

LEAD.

Acetate and Carb. Litharge Goulard's Ex.

Sulphate magnesium and phosphate sodium are both good antidotes for the soluble salts. For the solid forms, giving dilute sulphuric acid. The use of strychnia for the paralysis, and of iodide potassium, for the *chronic* forms generally, have been recommended.

MERCURY.

White and Red Precipitate. Cor. Sublimate Vermilion.

Albumen such as white of eggs, milk and wheat flour beaten with water, must be promptly administered. Counteract inflammation by ordinary means. Gold finely mixed in dust with iron filings. The iron filings and *ferri pulvis* have been given enclosed in gold leaf.

ZINC.

Acetate and Sulphate. White Vitriol.

The vomiting may be relieved by copious draughts of warm water. Carbonate sodium in solution will decompose the sulphate. Milk and albumen act as antidotes. General principles to be observed in the subsequent treatment.

CREASOTE.

Is immediately coagulated by albumen.

Phosphorus.

Matches, &c.

An emetic promptly; give copious draughts containing magnesia in suspension; mucilaginous drinks; general treatment for inflammatory symptoms.

ACRONARCOTIC AND NARCOTIC.

Aconite.
Baneberry.
Belladonna.
Bloodroot.
Calabar bean.
Camphor.
Cherry Laurel.
Coco. Indc.
Colchicum.
Curare.
Dogsbane.
Ergot.
Fox Glove.
Gelsemium.
Helebore.
Hemlock.
Henbane.
Lobelia.
Nux Vomica.
Opium.
Poison Oak.
Rue.
Squill.
Strammon.
Tobacco.
Verat. Vir.
Wild Cherry.
Wild Orange.

Evacuate the stomach with four or five grains of tarter emetic, or ten to twenty of sulphate zinc, repeated every quarter hour until the full effect is produced; assist by tickling the throat with a feather. Large and strong glysters of soap dissolved in water, or of salt and gruel, should be speedily administered to clear the bowels and assist in getting rid of the poison. Active purgatives may be given after vomiting has ceased. When as much as possible of the poison has been expelled, give alternately, a teacupful of strong hot coffee and diluted vinegar. If the drowsiness or insensibility be not relieved by these means, blood may be taken from the jugular vein, blisters applied to the neck and legs and the attention roused by every means possible. If the heat declines, warmth and frictions must be perseveringly used. Vegetable acid should on no account be given *before* the poison is expelled, and it is desirable that but little fluids of any kind should be administered.

93

WILLIAM R. WARNER & CO.'S

SUGAR-COATED

Tablet Triturates

(QUICKLY SOLUBLE)

Warner & Co.'s Sugar-Coated Triturates consist of the medicinal ingredient and pure Sugar of Milk, thoroughly mixed and incorporated together by trituration.

The ready solubility or diffusibility of the tablets in water and in the fluids of the stomach is worthy of the physician's attention.

Triturates, to be genuine preparations, must be made in accordance with certain well-defined principles.

They must have the precise quantity of medicament they are alleged to contain, and they must be accurate in dose.

The medicinal agent should be uniformly distributed through the whole mass.

	PER BOT. OF 500 EACH
Acid Arseniosum, 1-20 gr. - - - - $ .50
Acid Arseniosum, 1-30 gr. - - - - - .50
Acid Arseniosum, 1-40 gr. - - - - .50
Acid Arseniosum, 1-60 gr. - - - - - .50
Acid Arseniosum, 1-100 gr. - - - - .50
Aconitina, (Duquesnel's), 1-500 gr. - - .80
Aconitina, (Duquesnel's), 1-200 gr. - - 1.25
Aloin, 1-5 gr. - - - - - - .50
Aloin, 1-10 gr. - - - - - - .50
Calomel, 1-10 gr. - - - - - - .50
Cathart. Comp. Imp., 1-5 gr. - - - .50
Cathart. Comp. Imp., 1-10 gr. - - - .50
Cathart. Comp. U. S. P., 1-5 gr. - - - .50
Cathart. Comp. Veg., 1-5 gr. - - - .50
Corrosive Sublimate, 1-100 gr. - - - .50
Corrosive Sublimate, 1-200 gr. - - - .50
Ext. Nuc. Vomica, ¼ gr. - - - - .50
Ext. Colchici Fld., 1 min - - - - .50

WM. R. WARNER & CO.'S TABLET TRITURATES.

	PER BOT. OF 500 EACH
Iodide Mercury, 1-50 gr.	.50
Mercury with Chalk, ½ gr.	.50
Morphine Sulph., ¼ gr.	1.75
Morphine Sulph., ⅓ gr.	1.00
Morphine Sulph., ½ gr.	1.25
Podophyllin, ¼ gr.	.50
Podophyllin, 1-10 gr.	.50
Podophyllin, 1-20 gr.	.50
Salicinum, ½ gr.	.50
Strychnine, 1-100 gr.	.50
Strychnine, 1-200 gr.	.50
Tartar Emetic, 1-50 gr	.50
Tinct. Aconite, ¼ min.	.50
Tinct. Aconite, ½ min.	.50
Tinct. Aconite, 1 min.	.50
Tinct. Aconite, 2 min.	.50
Tinct. Belladonna, 1 min.	.50
Tinct. Belladonna, 3 min.	.50
Tinct. Belladonna, 5 min.	.50
Tinct. Cannabis Indica, 2 min	.50
Tinct. Cantharidis, 1 min.	.50
Tinct. Capsicum, 2 min.	.50
Tinct. Digitalis, 1 min.	.50
Tinct. Digitalis, 5 min.	.50
Tinct. Gelseminum, ½ gr.	.50
Tinct. Nucis Vomica, 1 min.	.50
Tinct. Nucis Vomica, 2 min	.50
Tinct. Nucis Vomica, 3 min	.50
Tinct. Pulsatilla, 2 min.	.50
Tinct. Strophanthus, 1 min.	.50
Tinct. Strophanthus, 2 min.	.50
Tinct. Veratri Viridis, ½ gr.	.50
Tinct. Veratri Viridis, 1 min.	.50
Tinct. Veratri Viridis, 2 min	.50
Tinct. Zingiberis, 3 min.	.50
Tinct. Zingiberis, 5 min.	.50

PREPARED BY

WM. R. WARNER & CO.

Manufacturing Chemists,

1228 Market St.,   18 Liberty St.,

PHILADELPHIA.   NEW YORK.

ELIXIR

SALICYLIC ACID COMP.

A POTENT AND RELIABLE REMEDY IN

RHEUMATISM, GOUT, LUMBAGO AND KINDRED COMPLAINTS.

This preparation combines, in a pleasant and agreeable form, Salicylic Acid, Cimicifuga, Gelseminum, Sodii Bi-Carb. and Potass. Iodid., so combined as to be more prompt and effective in the treatment of this class of diseases than either of the ingredients when administered alone.

This remedy can be given without producing any of the unpleasant results which so often follow the giving of Salicylic Acid and Salicylate Sodium—viz., gastric and intestinal irritation, nausea, delirium, deafness, nervous irritability, restlessness and rapid respiration; on the contrary, it gives prompt relief from pain and quiets the nerves without the aid of opiates.

The dose is from a tea to a dessertspoonful. Each teaspoonful contains five grains Salicylic Acid.

Elixir Salicylic Acid Comp. is put up in 12 oz. square bottles.

PREPARED BY

WILLIAM R. WARNER & CO.

MANUFACTURING CHEMISTS,

PHILADELPHIA and NEW YORK.

AN IMPORTANT NEW REMEDY.

LIQUID

PANCREOPEPSINE

—OR—

LIQUOR PANCREATICUS COMP.

(DIGESTIVE FLUID.)

This preparation contains in an agreeable form the natural and assimilative principles of the digestive fluid of the stomach, comprising **Pancreatine, Pepsin, Lactic and Muriatic Acids.** The best means of re-establishing digestion in enfeebled stomachs, where the power to assimilate and digest food is impaired, is to administer principles capable of communicating the elements necessary to convert food into nutriment.

The value of **Liquor Pancreopepsine** in this connection has been fully established, and we can recommend it with confidence to the profession as superior to pepsin alone. It aids in digesting animal and vegetable cooked food, fatty and amylaceous substances, and may be employed in all cases where from prolonged sickness or other causes, the alimentary processes are not in their normal condition.

It is usually given in tablespoonful doses after each meal, with an equal quantity of water or wine, or alone, as it is most pleasant and agreeable to the taste.

Put up in 16 oz. French Square Bottles
Price, $1.00.

PREPARED ONLY BY

WM. R. WARNER & CO.

PHILADELPHIA AND NEW YORK.

Paris  1889

Special and Private Recipes.

We solicit orders for your Special Recipe, and beg to say that our facilities for the manufacture of

SOLUBLE SUGAR AND GELATIN COATED

PILLS,

aided by extensive and improved machinery, enable us to furnish them at moderate prices. We are prepared to fill orders for millions of pills, but we cannot make less quantities than **3,000**, it being impracticable to sugar-coat a smaller number.

With a view to their proper manipulation, it is desirable to know the composition. We will therefore supply the ingredients and give the lowest estimate for same. None but the best and purest drugs are used in their manufacture. Our long experience and the favor with which our products are received, attest the excellence of our work.

Soliciting your orders, we are,

Very respectfully,

WM. R. WARNER & CO.

1228 Market St., Philadelphia.

Pills sent by mail, postage paid, on receipt of price.